Agents of the Current Thinking

CHRISTY LEPPANEN, PH.D.

Agents of the Current Thinking

ISBN: 979-8-9943616-7-2 (Paperback)

Library of Congress Control Number: 2025927973

First Edition: February 2026

Dedication

This book is dedicated to the hundreds of millions of people who—every day—use a product regulated by the U.S. Food and Drug Administration.

Acknowledgments

Thanks to family, old friends, and new friends. For your support and camaraderie discovered where I happened to plop my laptop. For conversations in coffee shops, hotel lobbies, and waiting rooms, on flights and at airports, parties, and parades, with so many offering to share their thoughts.

Thanks to former colleagues and to my former students. For your energy, wit, enthusiasm, and resolve to educate the masses
—to empower all individuals to make informed choices for themselves.

The collective message from you—your logic—propelled me to share this story. Regardless of our backgrounds or interests, we all want the truth.
We don't like being lied to.

Contents

Preface | The Art of the Deal

MY JOB AS A U.S. FEDERAL GOVERNMENT EMPLOYEE felt like a dream. In that dream, I was a gameshow contestant. Wayne Brady patiently held the microphone in front of me. The audience sat silently at the edge of their seats.

"Let's make a deal," he slowly said.

He then asked, "Keep Curtain #1? Or take Curtain #2?"

I knew what was behind Curtain #1. That was my current job. Behind that dingy green curtain were the workings of a failing federal agency tasked with regulating the sale of tobacco products while purportedly protecting the public health.

Curtain #2 offered an alternative.

On January 28th, 2025, an email with the subject "Fork in the Road" arrived from the U.S. Office of Personnel Management. The first of the new Trump administration's broad-brush efforts to reduce the size of government, it described the Deferred Resignation Program. It said, "If you resign under this program, you will retain all pay and benefits regardless of your daily workload and will be exempted from all applicable in-person work requirements until September 30th, 2025."

It said, "If you wish to resign: 1) Select 'Reply' to this email."

Then, "2) Type the word 'Resign' into the body of this reply email."

Then, "Hit 'Send.'"

What was really behind Curtain #2? There'd be no turning back and many warned "don't do it," that anyone who did would be left high and dry.

That January 28th, 2025, email also described several "pillars" that the new Trump administration would pursue, like a "Performance Culture" where "We will insist on excellence at every level—our performance standards will be updated to reward and promote those that exceed expectations and address in a fair and open way those who do not meet the high standards which the taxpayers of this country have a right to demand."

It described forthcoming "Enhanced standards of conduct" where "The federal workforce should be comprised of employees who are reliable, loyal, trustworthy, and who strive for excellence in their daily work."

It said, "Employees who engage in unlawful behavior or other misconduct will be prioritized for appropriate investigation and discipline, including termination."

That sounded great!

However, previous administrations said the same sorts of things, that there'd be improvements at the U.S. Food and Drug Administration (FDA). Those efforts and mechanisms designed to prevent and address wrongdoing, where I worked at the FDA's Center for Tobacco Products (CTP), proved to be performative. The Biden administration failed to correct pseudoscience worsened during the first Trump administration. Why would anyone think things would improve via Trump's second iteration?

So, Curtain #1 didn't seem like a good deal.

On January 30th, 2025, I typed "Resign," hit "Send," and drew the curtain on my job with the FDA. I could think of no better way to spend that time than to investigate, then share, what I'd been hired to disclose.

Fork it, dreamy Mr. Brady! I take Curtain #2.

1 | Fuzzy Math

IN APRIL OF 2023, I RECEIVED AN ASSIGNMENT to evaluate information about 16 combustible cigarette products. A tobacco company had requested 16 marketing orders that would allow those 16 products to be sold in the U.S. My evaluation would help satisfy CTP's obligation to implement the National Environmental Policy Act, or NEPA, the U.S. law requiring many federal actions be accompanied by a public communication to disclose information about exposures and impacts associated with those actions.

I'd first perform a review to determine if the information I had was sufficient to perform an environmental assessment for each of those 16 products. We used environmental assessments provided by applicants or wrote those assessments ourselves. Via those publicly posted assessments, the American people would be privy to the information CTP considered and how CTP made its determination that marketing each product in the U.S. was appropriate. CTP was also required to publicly post, for each tobacco product brand and subbrand, specific information about 93 harmful or potentially harmful constituents such as lead, arsenic, formaldehyde, and other carcinogens.

That transparency would help the American people make informed decisions about exposures associated with tobacco products, whether they chose to use them, were in spaces where others did or had, or wanted to inform and protect others. The environment that I was to consider included more natural environments like freshwater streams as well as the indoor air that we breathe.

The federal action to be considered was CTP's decision to grant a marketing order. The Family Smoking Prevention and Tobacco Control Act of 2009 required that CTP determine if concerns about a product should preclude its sale. If not, the product would receive a marketing order and could be sold.

That said, many products were sold for years in the U.S. without marketing orders and without undergoing a regulatory review. That exposed users to any known or undescribed hazards from those products. Users and non-users would be exposed via any associated means that contaminated their surroundings. Contaminants emitted from cigarettes are known to cause death and disease, and amounts that accumulate indoors are harmful. Aerosols from electronic nicotine delivery systems (ENDS), commonly known as vapes, also contaminate our air and surroundings. Exposures come from the use of all product types as well as from product and packaging disposal and degradation.

Alarmingly, illicit products that constituted the majority of some U.S. markets altogether eluded CTP's required regulatory review. CTP allowed millions of untested illicit ENDS products, oral pouch products, and combustible tobacco products to flood the U.S. market. For example, on November 6th, 2024, the Truth Initiative reported that illicit products constituted an estimated 86% of the ENDS market. And CTP knew many of those products appealed to youth that, compared with older first-time users, might be more likely to become lifelong nicotine addicts.

Anyway, my training indicated our assessments needed to be informed. However, I faced a problem that prevented me evaluating each of those 16 products I'd been assigned to evaluate. I needed to consider exposures associated with each product's manufacturing, use, and disposal, the standard things we were to evaluate. However, application materials submitted by the applicant seeking those marketing orders included marketing estimates, or the likely number of units sold in the U.S., not for *each* of those 16 products but *combined* for 32 products. Information about the 16 products that were the focus of my assignment was combined with information about 16 other products that I wasn't evaluating. Other reviewers faced the same or a similar scenario in their assignments to evaluate products from the same applicant.

A simple solution would've been to ask the applicant to parse marketing estimates for each of those products. Instead, we were directed to lie.

2 | The Setting

I'D BEEN A TOXICOLOGIST, biologist, and primary reviewer of applications for marketing orders for tobacco products for a bit over two years when, in May of 2023, we were directed to lie. I worked in the Division of NonClinical Science, in CTP's Office of Science that supported the scientific rationale behind decisions to grant or deny those marketing orders. My role was to support CTP's regulatory decision making and reporting requirements. So, I evaluated all relevant information about tobacco products considered for sale in the U.S.

When a company wanted to sell a new tobacco product in the U.S., it needed to apply for a marketing order that allowed the legal sale of that product. For instance, the applicant wishing to sell those 16 products paid a fee with its application for a marketing order for each of those 16 products. Those fees funded CTP and paid my salary. Because our work was funded by fees, things like government shutdowns wouldn't delay marketing orders. We kept working with pay. CTP processed those applications, and we worked on a timeline.

After a company submitted its application, it received a letter from CTP stating that its application had been received. CTP then performed an acceptance review. If the type of information needed to perform a scientific review had been provided or was available, the applicant then received a letter from CTP stating that the product had been accepted for scientific review. Those 16 products were submitted, accepted for review, and reviewed under CTP's Exemption Request (EX REQ) pathway.

Different scientific review pathways had different timelines before the agency's decision was to be communicated. An applicant could request an exemption from substantive scientific review via the EX REQ pathway, with an expected 60-day turnaround. The EX REQ pathway was for new products developed as an outcome of increasing, decreasing, adding, or deleting additives from an already marketed comparator product. The Substantial Equivalence (SE) pathway was for products with the same or similar characteristics as a comparator product and a response was expected in 90 days. Products accepted via EX REQ and SE pathways were considered altered versions of their pre-existing comparator products, and information about those pre-existing comparator products helped inform regulatory reviews of those new products.

The Premarket Tobacco Product Application (PMTA) pathway also involved comparisons with available comparator products. Because PMTA products were sometimes less aligned with what had come before, very extensive scientific reviews were needed, and turnarounds were expected in 180 days.

The Modified Risk Tobacco Product Application (MRTPA) pathway considered products that might be designated less harmful, a big claim that could be used to market a product. Therefore, those reviews typically took the longest, for example, owing to public disclosures and comments that the agency would need to address. Any modified risk designation therefore followed a very, very extensive scientific review.

CTP received $712 million annually and, as of March 2nd, 2025, employed about 1,200 employees in six offices: the Office of the Center Director, Office of Management, Office of Regulations, Office of Health Communication and Education, Office of Compliance and Enforcement, and the Office of Science. CTP's organization generally constituted about five to ten employees in a team and three teams in a branch. Each team was supervised by a team supervisor. Each branch was supervised by a branch chief. In general, each division was made up of several branches and each office of several divisions. And each division and each office had its own set of leadership staff. CTP was housed in the FDA, that was overseen by the U.S. Department of Health and Human Services (HHS).

Throughout the accounts to follow, I've edited for clarity and consistency without altering meaning or mood. For example, terms are standardized, many abbreviations are spelled out, and redundant language is mostly excluded. However, apparent misspeaks and typos are mostly maintained. The names of people, products, and the producers of products discussed are mostly redacted. The requirement that I not disclose business information that is confidential and not publicly available persists even after my departure from federal employment. But my oath will supersede any pressure to hide institutional indiscretions.

Changes in leadership during my four years at CTP, from February of 2021 to March of 2025, are reflected with a number after each position title that represents consecutive staff changes. If no number follows a title, one individual consistently served in that role, as did our Deputy Director for Regulatory

Management, Division Director, and Division Associate Director. On the other hand, in 2023 and 2024, the Environmental Science Branch where I worked was first overseen by Branch Chief #1, then by acting Branch Chief #2, then by acting Branch Chief #3, then by acting Branch Chief #4, then by acting Branch Chief #5, then by acting Branch Chief #6, and finally by Branch Chief #7. Our branch was made up of three teams, and I was on Team 1. In 2024 alone, Team 1 was supervised by Supervisor #2, then by acting Supervisor #3, then by acting Supervisor #4, then by acting Supervisor #5, then by acting Supervisor #6, and then by acting Supervisor #7.

The aim of using titles as such, capitalized when they replace the proper name of a specific individual, is to focus on the bureaucracy by simplifying staff changes that are nonetheless not inconsequential. In just a few years, my branch doubled to over two dozen employees and my division from around 50 to more than 100, despite considerable attrition. During my time at CTP, many leadership left and were replaced following many months when temporary staff rotated through those roles. Individuals with different backgrounds and disturbingly different takes on how we implement the law moved from one position to another for sometimes less than a few weeks. Some popped in from outside our branch or moved among many leadership roles.

Most individuals holding various positions are usually referenced by their most influential role. For example, our Division Associate Director, who was also our Branch Chief #2 and #6 and our Team 1 Supervisor #3, is referenced as our Division Associate Director. The Team 2 Supervisor #1 and #3, who was also our Branch Chief #3 and #5, is referenced as Branch Chief #3 and #5. The Team 3 Supervisor #1 and #3, who was also our Branch Chief #4, is referenced as Branch Chief #4. The Team 2 Supervisor #2, who was also the Team 3 Supervisor #2 and became our Branch Chief #7, is referenced as Branch Chief #7. And the Team 3 Supervisor #4 who became my Team 1 Supervisor #7 is referenced as Supervisor #7.

Because I was on Team 1 and talk mostly about Team 1, any supervisor without a team number is my Team 1 supervisor. I also worked with various secondary reviewers and numbered three of them involved with the representative accounts provided: Secondary Reviewer #1, Secondary Reviewer #2, and Secondary Reviewer #3.

If you're confused, I've established a good foundation for what's to come. Who these people are and what they did matters, but I'm not sure how we hold them accountable, and we need to focus just as much on correcting faults in the system that allowed them to choose the behaviors they chose.

Leadership Oversight at CTP, FDA, and HHS
March 2nd, 2025

The **Center for Tobacco Products** Director oversaw about 1,200 employees:

Office of the Center Director
Office of Management
Office of Regulations
Office of Science
Office of Health Communication and Education
Office of Compliance and Enforcement

The **Food and Drug Administration** Commissioner oversaw 20,128 employees:

Office of the Commissioner
Office of the Chief Counsel
Office of the Executive Secretariat
Office of the Counselor to the Commissioner
Office of Digital Transformation
Office of Enterprise Transformation
Office of Planning, Evaluation & Risk Management
Center for Biologics Evaluation & Research
Center for Devices & Radiological Health
Center for Drug Evaluation & Research
Center for Tobacco Products
Center for Veterinary Medicine
Oncology Center of Excellence
Office of the Chief Medical Officer
Office of External Affairs
Office of Minority Health & Health Equity
Office of Operations
Office of Policy, Legislation, & International Affairs
Office of the Chief Scientist
Office of Women's Health
Human Foods Program
Office of Inspections & Investigations

The **U.S. Department of Health and Human Services** Secretary oversaw about 88,000 employees:

- Deputy Secretary
- Chief of Staff
- Immediate Office of the Secretary (IOS)
- Office of Intergovernmental and External Affairs (IAE)
- Office of the Secretary
    - Office of the Assistant Secretary for Administration (ASA)
    - Office of the Assistant Secretary for Financial Resources (ASFR)
    - Office of the Assistant Secretary for Health (OASH)
    - Office of the Assistant Secretary for Legislation (ASL)
    - Office of the Assistant Secretary for Planning and Evaluation (ASPE)
    - Office of the Assistant Secretary for Public Affairs (ASPA)
    - Office for Civil Rights (OCR)
    - Departmental Appeals Board (DAB)
    - Office of the General Counsel (OGC)
    - Office of Global Affairs (OGA)
    - Office of Inspector General (OIG)
    - Office of Medicare Hearings and Appeals (OMHA)
    - Office of the Assistant Secretary for Technology Policy and the Office of the National Coordinator for Health Information Technology (ASTP/ONC)
    - HHS Chief Information Officer
- Operating Divisions
    - Administration for Children and Families (ACF)
    - Administration for Community Living (ACL)
    - Agency for Healthcare Research and Quality (AHRQ)
    - Advanced Research Projects Agency for Health (ARPA-H)
    - Administration for Strategic Preparedness and Response (ASPR)
    - Agency for Toxic Substances and Disease Registry (ATSDR)
    - Centers for Disease Control and Prevention (CDC)
    - Centers for Medicare & Medicaid Services (CMS)
    - Food and Drug Administration (FDA)
    - Health Resources and Services Administration (HRSA)
    - Indian Health Service (IHS)
    - National Institutes of Health (NIH)
    - Substance Abuse & Mental Health Services Administration (SAMHSA)

3 | Covering Our Butts

ALTHOUGH IT WAS A STRETCH to consider our regulatory work scientific, I'd never been directed to outright lie. It was astonishing that the directive came in writing from our Branch Chief #4. That directive was supported in writing by our Branch Chief #3 and #5 and by a longtime reviewer who wrote that it was a means "to cover our butts."

We were to include throughout the part of our assessments that the public would see something akin to "The agency based its determination on marketing estimates." Marketing estimates would then be included in a table in a confidential appendix that the public wouldn't see. Including that information in a confidential appendix wasn't odd because marketing estimates were considered confidential business information. To "cover our butts" with respect to the fact that data pooled for 32 products didn't inform individual considerations of 16 different products, we were directed to add in small text below that table in that confidential appendix something akin to "The marketing estimates constitute estimates for all new products."

Anyone reading that wouldn't understand what that meant. There was no indication that those estimates constituted a sum of marketing estimates for 32 new products or that any calculations or representations of that pooled data represented 32 products. So, even if the public could see the entire document and all of the information in that confidential appendix, the approach created the perception that marketing estimates applied to 16 products. And it was explicit in what the applicant provided that marketing estimates were pooled for 32 products, but the public couldn't see the applicant's submission.

We'd sometimes use assessments that the applicant provided, posting those documents publicly instead, because that was more efficient. In this case, we'd instead recreate the documents they'd provided, reworked with data misrepresented and deceptive arguments added.

For example, one assessment written by a colleague included in the public facing part of the document:

The agency evaluated potential impacts to resources in the environment that may be affected by use of the new products and found no significant impacts based on agency-gathered information and the applicant's submitted information. Included in the information the agency considered were the projected marketing estimates for the first- and fifth years of marketing the new products in the United States (Confidential Appendix 2).

Then in small text in a footnote at the bottom of Confidential Appendix 2 it says, "Marketing estimate depicts the total of all new products and their modifications in [a truncated product tracking number]." In another assessment, the footnote says, "Values represent the total projected marketing estimate for all the different versions of the new product," then likewise references a truncated tracking number. In the redacted tables above each footnote, the marketing estimate for 32 products is listed next to the list of 16 products with headings that create the impression that the number for 32 products represents the list of 16.

The excluded part of those truncated tracking numbers, known as Submission Tracking Numbers or STNs, would've indicated there are more than 16 new products. Hypothetically, say the 32 tracking numbers ranged from EX999999.PD1 to EX999999.PD32, where EX refers to the EX REQ application pathway and PD to the product number. Our training, that explicitly directed us to include those PD numbers, meant the footnotes should say, "EX999999.PD1-32." Those deceptive footnotes only say, "EX999999."

This was an intentionally orchestrated set of lies. Its application was argued for weeks among our three teams and with branch leadership. My Supervisor #2, who was also my Secondary Reviewer #1 on that assignment, eventually offered primary reviewers the option to altogether exclude marketing estimates and the deceptive language. I did that. However, without marketing estimates, it was then unclear how the agency would reach its determination.

4 | Freeform Fallacies

MARKETING ESTIMATES, IN MY EXPERIENCE, were something we'd always included in our assessments. We pointed to marketing estimates, provided in application materials from applicants seeking marketing orders, as informing our agency's decisions. We included in our assessment of a new product the number of units for the new product and for its comparator product referred to as a *pre-existing*, *predicate*, or *original* product. We included that information for the current year and for the first and fifth years after marketing orders would be granted.

Previously, in my experience, we'd always asked applicants to provide marketing estimates that were missing or to clear up any confusion about marketing estimates they'd submitted. I recall no instance where a queried applicant hadn't provided that information or indicated they couldn't, for example, because they didn't manufacture or market the comparator product.

Our Division Director, my Supervisors #2 and #4, and others explained that the agency determined there were no significant negative impacts of marketing a tobacco product, and therefore marketing orders were facilitated, because the same number or fewer units of a given new product would replace a comparator product that would be removed from the market. This was, at most, a one-to-one product replacement, for example, where no more than one new cigarette would replace one previously marketed cigarette. Therefore, in practice, a determination of no significant impacts would be made, for example, if 100,000 units of a new product replaced 100,000 units of a similar comparator product.

The rationale was that if no *additional* products go on the market, there'd be no additional impacts. The marketing order could allow the same amounts of the same toxicants, like lead, arsenic, formaldehyde and other carcinogens, resulting in the same amount of death and disease. In this context, CTP's granting of marketing orders would sicken and kill the same number of people, just not any additional people, and that protected the public health.

However, because that rationale often didn't apply, staff frequently asked, "Are we [still] assuming a one-to-one replacement with the predicate or original products?"

"Yes," they'd say we still assumed that.

And they said we determined there were no significant impacts based on a one-to-one replacement in a market that was not increasing. However, sometimes it was, and there often wasn't a one-to-one replacement.

For simplicity's sake, let's assume that halving or doubling the number of products halves or doubles, respectively, any associated impacts. That might not be the case, for example, if a greater proportion of products are unused or littered when more products are sold. A larger market might also be associated with a higher incidence of addiction, disease, and death that would therefore come with greater associated knock-on effects.

A marketing estimate provided by an applicant for a new product they hoped to sell was often less than the estimate they provided for the comparator product that the new product would replace. For example, 50,000 units of a new product replaced 100,000 units of a comparator product, cutting impacts in half. And no doubt such a scenario could be legitimate. However, a new product's marketing estimate was sometimes greater than that of the comparator product it replaced. For instance, 200,000 units replaced 100,000 units, doubling any impacts.

Additionally, one comparator product could be used as a comparator with many new products. Imagine 20 new products each marketed at 50,000 units each, each associated with the same comparator product previously marketed at 100,000 units that would no longer be marketed. The combined new product marketing estimate for the 20 new products of 1,000,000 units would exceed the 100,000 units associated with the comparator product, increasing impacts by a factor of 10. An applicant or many applicants could repeatedly choose the same comparator product, one that they produced or one produced by another company. And different companies wouldn't necessarily know they'd used the same comparator product.

These sorts of scenarios weren't infrequent. And we could've easily done it, but tracking repeated use of a comparator product wasn't allowed. We asked. Each product was assigned a unique Submission Tracking Number. Entering a comparator product's STN into our databases would have returned all associated new products as well as documents that contained each of those products' marketing estimates.

Additionally, comparator products often weren't removed from the market. Considering the scenario above, marketing the 20 new products *with* the comparator product that wouldn't be removed from the market increases the market from 100,000 to 1,100,000 units.

Considerations of marketing estimates often constituted the only quantitative argument associated with the new product under consideration that the agency used to reach its findings. Other data we retrieved from public databases constituted manufacturing discharges associated with a facility that produced many products, but that information wasn't available for all facilities or by product. And we made no attempt to confirm or validate in any way that information, that had been provided by an industry still known for deceptive practices. We didn't ask for any information or explanation about how those estimates were developed. We didn't ask or try to determine if marketing estimates provided by the industry were subsequently realized. And we certainly could've done those things as well.

Further, when we used market trend data to indicate sales were decreasing, and therefore a new product wasn't associated with any new problems because it entered a market in decline, we based that, again, on data that was self-reported by tobacco companies. CTP paid for access to other data, for example from Nielsen and Euromonitor International, that instead sometimes showed market increases that we ignored. And when increases became apparent in even the industry-reported data, we stopped including those trends in our assessments. We didn't apply the same rationale and indicate that adding a new product to an increasing market might suggest an additional negative outcome.

In many cases, marketing estimates provided by applicants made it impossible for the agency to abide by its own scientific integrity policies and the one-to-one product replacement rationale that our leadership confirmed was the basis of the agency's determinations. And because marketing estimates are considered confidential business information, that information and therefore the basis of each agency decision was hidden from view in redacted confidential appendices. Looking online at our publicly posted documents, it struck me how redactions might support the false assumption of something substantive.

5 | *In*credibly Consistent

I'D LEFT FOR A VACATION right after I submitted my assessments for those 16 combustible cigarette products that excluded marketing estimates and those deceptive arguments. I was relieved that I didn't need to lie. That said, what I'd done and the less blatantly fraudulent approaches we used didn't feel like the opposite of a lie. And when I returned, I saw the deceptive approach had still been used in some assessments. One other reviewer besides me had altogether removed marketing estimates, but others hadn't. So, those lies had now been entered into the record.

I was told that we needed to be consistent, that assessments produced by all staff should reflect the same approaches and cite the same references. Therefore, we needed to use the same representations of marketing estimates.

It made sense that we be consistent because the same science and same sound approaches should be applied via the same means across all products. Fees paid by all importers and domestic manufacturers of cigarettes, cigars, snuff, and pipe, chewing, and roll-your-own tobacco funded the process by which CTP denied or granted marketing orders. Fees weren't required for ENDS products, dissolvables, or nicotine gels. However, because our mission was to protect the public health, all products should've been similarly considered.

Consistency ensured a few things. It ensured all products received the same level of scrutiny before going on the market. This meant the American people could be comfortable the products they used had passed through a consistent set of filters. And reasons products were denied marketing orders should also be consistent. Regardless of anyone's opinion about tobacco products, those abiding by U.S. law that apply in good faith for marketing orders also deserve consistency.

However, consistent use of the same approaches risked excluding information needed to protect the public health and needed to comply with the laws we were obliged to implement. I'd also been told that I was responsible for defending the approaches that I used in the reviews and assessments that I prepared and signed. Our regulatory work also needed to be consistent with the current science. Therefore, I couldn't simply cut and paste the work of others.

That also made sense and would collectively help inform our work. Different reviewers with different backgrounds and experience would know and find different things that were applicable and important. New data, models, approaches, and interpretations of science and the law would need to be considered. Any public health agency tasked with applying sound science should certainly include new considerations of new science, or any sound approaches not previously considered, as a mainstay of its processes.

Some primary reviewers said they were uncomfortable further raising concerns about directives to use the deceptive approach. One said she was afraid because she hadn't completed her probationary period as a federal employee, a period during which an employee can more easily be fired. Another said he didn't want to cause trouble for his newly appointed supervisor. Regardless, his Team 3 Supervisor #2 with all our branch leadership knew about the controversy but still approved the approach and signed those documents. These documents went through their filters to ensure our work was credible yet remained consistent, two things that couldn't coexist.

6 | "This is not tough stuff, like, it's math. It's simple math."

IN DECEMBER OF 2022, NEW MANDATORY TRAINING directed staff with concerns about scientific integrity to our CTP Office of the Ombudsman. So, that's what I did. That office was responsible for implementing a new Scientific Dispute Resolution (SDR) process and I followed that process. I provided written proof of the deceptive approach with written proof that scientists were directed to lie when developing and implementing that approach.

The CTP Ombudsman and Associate Ombudsman determined those issues qualified to be heard under that new SDR process. They believed the issues were critical and that unsuccessful efforts to address those behaviors and directives, from April to August of 2023, warranted escalation.

Our Branch Chief #4 had written that he didn't "believe" it was inaccurate to include disinformation in our assessments. However, we were required to address inconsistencies in our regulatory review work, efforts to promote public disinformation, and pressure on scientists to lie. My Supervisor #2 said letting two of us exclude marketing estimates solved the problem and refused to discuss violations of scientific integrity. He often wouldn't respond in writing or sat in silence. At first, I'd break the silence but eventually also sat quietly for sometimes more than 30 seconds. He'd then repeat some derivation of "They won't let us ask for that." If I asked who "they" were, he'd again say "they" or "leadership" or "Office of Science leadership," and sometimes said he didn't know.

The CTP Ombudsman and Associate Ombudsman offered to help frame the issues, schedule and attend meetings, and navigate a solution. They said I was brave to bring issues forward, that doing that was important but that few employees would do it. They said this was the first time the new SDR process would be used and we all seemed excited to put that new resource to use.

I prepared a draft email with accounts of the deceptive approach. They then provided suggested edits that I accepted. Then at 12:46 pm on August 1st, 2023, I sent that email with evidence in attachments to my Supervisor #2 and to our Division Associate Director. I requested a private meeting.

Our rotation of branch chiefs, who supervised our supervisors and wouldn't engage in good faith, were supervised by our Division Associate Director. He oversaw our regulatory reviews and assessments. After he reviewed our assessments, he signed determinations for the agency about exposures and impacts that might require additional considerations before marketing orders could be granted.

He responded on the same day at 12:59 pm, and he copied our Branch Chief #3 and #5 and others. He wrote, "It will take some time for me to read through all of this and under the priorities given to me. Please cc your Acting Branch Chief. Based on your email it seems you have a lot of concern so I'm ccing [our Division Director]. However, he is out of office and [another supervisor] is Acting for me. I appreciate you communicating your views and any concerns or ideas."

He'd always engaged before. But now, only 13 minutes after that email was sent, he'd determined what he'd seen wasn't enough of a priority to address in any way any time soon. What he'd seen was succinct and included organized evidence. We'd intentionally prepared that email for easy digestion.

The first sentence in my email, that stood alone with two hard returns after it, said, "I am requesting a time to privately speak with you both regarding concerns with the ongoing practice of creating and entering into the record inaccurate documents by the CTP Office of Science, Division of NonClinical Science, Environmental Science Branch."

Directly to follow that statement, I'd written, "As I understand my job, it is to support FDA's public health mission by creating accurate, credible, and appropriate work products that withstand scrutiny. Recently, when nine reviewers were each assigned a bundle of similar products from the same applicant, in working together, a questionable practice came to light. Reviewers were directed to pretend we had applicable marketing estimates when we did not."

Clear evidence followed.

I'd asked to meet privately with him and my Supervisor #2, and the CTP Ombudsman and/or Associate Ombudsman would also be present. The CTP Ombudsman said raising workplace issues was sometimes associated with retaliation. He suggested I request a private meeting with these two individuals because our Branch Chief #4 and Branch Chief #3 and #5 were complicit. The email and evidence provided to the Division Associate Director clearly indicated their involvement and clearly requested a private meeting. However, he'd copied Branch Chief #3 and #5 and others in his response.

He never circled back, and my Supervisor #2 didn't respond. None of them responded or cooperated.

I did respond to him:

I understand that people are busy, so I'll keep an ongoing list of concerns and ideas and look forward to your questions when we talk. In short, in my opinion, we need guidance, accountability, and consistency. One idea is to share in our branch how issues raised during the review process are addressed, even if those concerns are not communicated to the applicant. Our standard should not be predecisional, e.g., I'm concerned if one day we identify an environmental impact that the record we've created will prevent us from addressing it. Identifying the problems alongside positive solutions that improve the credibility of our work products might go a long way.

Anticipating future engagement about my views, concerns, and ideas, I needed to prepare. I could certainly go over what had happened alongside the accounts and evidence already provided. However, "communicating [my] views and any concerns or ideas," indicated I suggest possible solutions and required more insight. With respect to the deceptive approach, an easy solution was that we ask the applicant to parse marketing estimates by product. But I needed to give this more thought. I needed to try to understand why this had happened.

I was reminded about a few things our Branch Chief #4 had written months earlier in May of 2023 when he'd directed us to use the deceptive approach. With respect to our need for marketing estimates for each product, he wrote, "The public doesn't see these numbers anyway because that information is confidential, so we're not telling the public anything by asking for that information."

With that information our assessment would be *informed*, regardless of whether the public would recognize any difference. The deceptive assessments included text indicating that the agency had considered and made decisions using information the agency didn't have. Requesting that information would make those assessments honest if the information we received supported the same conclusions. If it didn't, we could correct those assessments.

He wrote, "Also, if we ask now, we not only have to face issues with SPB but also with [the tobacco company]'s legal team since we haven't asked this for the last 30 applications they've sent in." Further, he wrote, "Also please realize that if we change our tactics with these bundles after reviewing so many, we'll need to have VERY good reasons as to why and we'll be rightly challenged by our own leadership as well as face quite a backlash from applicants."

He often spoke dramatically about internal conflict, about what we'd face from boogeymen or flying monkeys, in the context of important professional interactions with which our training indicated we should engage. I had no fear of describing reasons that aligned with agency training and messaging.

And yikes! The deceptive approach had been used before!

We were expected to continue fuzzy math to prevent a lawsuit when a change to more credible work would be discovered by tobacco company lawyers patrolling for mistakes. This meant we'd all need to continue and sign our names to lies devised by others. We couldn't change tactics to create a credible work product. And we were also keeping our own legal experts, SPB, behind a smokescreen. Our Science Policy Branch would normally be looped in about anything like this.

I'd been concerned about what entering the deceptive approach into the federal record might do to our credibility, but it was already there!

Looking into the record, I found several instances where the deceptive approach had been used before. I'd been away from my FDA job, deployed on a humanitarian mission at the Southwest border with the Department of Homeland Security in February and March of 2023, so I'd missed those discussions. That said, review-related challenges and unique scenarios were to be shared with all staff during branch meetings or via other communications. But the practice hadn't been captured in meeting notes or via other means, not during the previous iteration or other instances when the deceptive approach had been used.

I could see in the record that all leadership had previously signed off on the deceptive approach. We'd initially been concerned that our Division Associate Director was in the dark about what was happening in our branch, and we didn't want him to be blindsided. However, our Division Associate Director, all Branch Chiefs #2 through #7, and all team supervisors—all permanent and acting leadership including my Supervisor #2—had previously reviewed documents that included the deceptive approach, indicated their concurrence with what was included in those documents, signed those documents, and allowed many instances of the deceptive approach to be entered into the federal record.

When I'd said we needed to prevent the practice from moving into the federal record, no one bothered saying it was already there. My Supervisor #2 sat in silence or acted oblivious, knowing, while I raised that concern for months, that CTP had done this before, according to our Branch Chief #4, at least 30 times that involved many combustible cigarette products each time.

Yikes.

So, initially, we were all expected to use the deceptive approach because others already had. Then, when examining the issue revealed that it wasn't up to snuff, I and another reviewer were offered the ability to exclude the deceptive approach from *our* assessments. But at the same time, other reviewers continued using it, some even after we'd completed our assessments that excluded it.

My Supervisor #2 was one of three team supervisors who supervised all reviewers in our branch. Those team supervisors worked together to ensure our work products were consistent, and they met a lot with respect to this issue. So, why hadn't our work products been consistent? Had Supervisor #2 offered the more aboveboard approach to placate us? Had we been *handled* with hopes that no one would notice the difference?

Several weeks passed with no follow-up from Supervisor #2 or our Division Associate Director, or from others he'd looped in like our Division Director. So, the CTP Ombudsman said it was time to raise the issues another rung in the hierarchy. The same set of accounts and evidence was therefore emailed with a meeting request to our new Office of Science Director #2. He'd been on board for about five months, and right away scheduled a meeting for August 29th, 2023. Our previous Office of Science Director #1 left in July of 2022 without pause to work for a tobacco company. In the interim, these sorts of issues had been handled by the Office of Science's Deputy Director for Regulatory Management.

Before we met, our Division Director sent me an instant message, "Do you have a moment?" that on August 28th, 2023, was followed by a quick call that was a little odd. He said, "I just have *one* question for you I need to get out of the way because you're going to have a conversation with [our Office of Science Director #2]. I just want to ask *one* question: Did [our Division Associate Director] talk to you at all about the issues you are raising?"

He hadn't, so I said "No."

"I just have that." he abruptly said. "Okay," he said, "That was all. Okay. Have a *wonderful* day. That's all I got."

"Okay," I said, "you too."

Our CTP Ombudsman, CTP Associate Ombudsman, and I then met with our Office of Science Director #2 on August 29th, 2023. He'd reviewed the evidence, but I'd additionally developed, shared, and presented slides that demonstrated what had happened, referencing written evidence throughout. He said we shouldn't be using the deceptive approach, said it needed to be corrected, and planned a meeting to develop a solution.

We talked about marketing estimates, in general. He asked, "Do they explain how they get the one year and five years post-marketing granted orders estimates? Because, for example, the products have been around for a while. They would have trend data, and they would know whether the sales for that product was going up or going down or staying steady and whether or not, you know, whatever the changes are."

"They don't explain to us how they derive their information," I said. "They don't tell us how they come up with these numbers. And should we ask them that?" I said, "Economists in CTP use Euromonitor International data to look at trend information in different product categories. We noticed it showing an increase in cigar trends over time. Whereas we use Treasury information, and that information shows a decrease over time. We need to be careful about this because CTP is using information that's contradictory, and it looks in both cases to serve a purpose. Right? So, the trend information that shows an increase in sales in Euromonitor information is used to support a concern about public health. Whereas the Treasury information we use in our assessments shows a decrease, and that's used to find a finding of no significant impact. So, that's contradictory within CTP."

He said, "You can look at historic trends for an individual UPC, you know, for a specific product. I started analyzing Nielsen data in 1997 with the Office on Smoking and Health. I was actually the one that introduced the notion of analyzing Nielsen data a million years ago. So, I'm really familiar with this. I'm a health economist. I've analyzed sales data my whole career. So, that's why I'm asking a bit more than your average bear."

He continued, "But anyway, I mean we're sort of deviating, but I think, I do think it's reasonable to ask for how they get those projections, because, you know they, their bread and butter is, of course, sales. And so, I'm sure they have very good data on their own sales." He said, "I'm sure [the applicant] knows exactly what's going on with their sales data and is probably making projections, because those projections are important to them." He said, "I think it's reasonable to ask to justify the numbers. And it's probably not even that difficult for them to comply to that."

He continued, "So, anyway, *this* is a more basic math problem. If you report on 32 products and you're really looking at 16, you obviously can't really draw strong conclusions."

"I appreciate this discussion," and I said, "I felt sort of silly putting this together in this much detail because that's the gist of it. It is that we have information for 32 products that is being represented as 16."

He said, "I can imagine how someone rationalizes it." He said, "They could argue that, well, the 32, the trends in the 32 are probably, you know, not going to be, are going to be fairly representative of the 16 as well, because maybe they're all moving together. But you're having to make assumption, you know. It could be unlikely, but it could be one set of 16 is going down a lot and then the other is going up a lot. And then that effect is nothing. And of course you can't tell that if it's all mushed together. So, this just seems like a pretty basic problem and should be called out." He said, "I can't see a reason why not to push back and make this deficiency," meaning that the data were *deficient* for our needs, and we'd ask for data parsed by product.

He said about branch leadership that brushed off our need to do credible work, "It seems like we're, we have to be a bit, in our job, we have to be more *exacting*. And the burden of reporting this accurately, to me, as *trivial*, just seems sloppy to me."

I said, "We're creating a record that can't be defended because these decisions are intentional. And it's intentional in the record. So, we have Microsoft Teams chats, emails, but—if [the applicant] did a Freedom of Information Act request, I think, you know. One of our acting branch chiefs said, 'Well, nobody can see it, so it doesn't matter,' and that's not really true."

"No," he agreed. He said, "I will say, even from my brief tenure, uh, it's naïve to think that things that are emailed, chatted about or whatever, don't eventually see the light of day. Most things see the light of day, um, when an issue gets, you know, *scrutiny*. So, um, you know, the way that we should be working in terms of doing our science is that it stands the scrutiny of outside inspection. Even if it isn't currently, we should be acting as though it is, that it should be *defensible*, um, you know. And any time people are having disagreements that they can't get resolved, that it should get escalated."

He said, "I know a whole lot about sales data and market volumes and the level of detail." He said, "The applicant needs to correct it to match the records, so that in the case of the 16, they just need to revise that and provide their data that matches the 16. I just don't see that as, you know, like an undue burden, especially compared to all the other things that are done."

He said, "We need to be explicit about this and just make decisions so that there's a memo to make it clear, so that there is consistency and it's accurate."

He said, "What we just need is some leadership on this and a memo, I think, to just clean it up," and "identify, you know, somebody to draft the guidance so we can just put this to bed and make sure that we're being consistent."

He said, "[Our Division Director] had responded to me quickly, to your message that you forwarded, and his, he said essentially the same thing, that we just need a memo on this."

He continued, "My take on this is, I send an email to [our Division Director] and [Division Associate Director] and copy Christy and say, like, 'Let's just meet briefly and talk about a process to set up simple guidance for this.' Like, to me, this is not tough stuff. Like, it's math. It's simple math, actually, um, you know."

At the end of the call, the CTP Ombudsman said to me, "Thank you for bringing it to our attention. I know it takes a lot of courage to, you know, point things out and go against the *common denominators*. So, thank you for your honesty and your strength in this."

Our Office of Science Director #2 added, "Yeah, and it's definitely what I hope for in the Office of Science, that, you know, I have an open-door policy, and if people see that things are just not taken care of, you know, I don't mind. I don't mind having conversations like this. Sometimes the solutions are difficult. Sometimes they're easy. But, you know, if it's causing heartache and disagreement, we just need to nip it in the bud and deal with it."

On August 31st, 2023, our Office of Science Director #2 and I then met with the Division Associate Director and the Division Director. The Division Associate Director summarized the deceptive approach, and we discussed some of the other problematic issues associated with our use of marketing estimates. Our Office of Science Director #2 indicated, he said based on his over 17 years of experience with tobacco industry data, that marketing estimates for each individual product could have easily been provided, if we'd asked. He said, moving forward, we'd ask the applicant to parse marketing estimates by product.

Success!

That new approach, that precluded any future use of the deceptive approach, was subsequently implemented. However, moving forward we were to ask applicants how they derived their information, but we didn't. And assessments where the deceptive approach had already been used weren't addressed. We hadn't addressed concerns about the federal record or about leadership pressuring scientists to lie. I reiterated those and other critical concerns to them, as I was obliged to do, and waited.

During that August 31st, 2023, meeting, our Division Director also said we'd develop a memo to clarify how marketing estimates should and should not be used. However, no clarification was developed. Instead, branch leadership was uncooperative and even ended related, ongoing efforts. Subsequent, repeated written directives from our CTP Director #2 to develop that clarification were likewise ignored.

7 | "There's just like, a very like, Wizard of Oz like man behind the curtain kind of like world that goes on."

WE WEREN'T JUST TALKING ABOUT HOW TO CORRECT one issue or correct or amend the record with respect to that deceptive approach. Leadership behavior was also a concern. That's how those problems came about in the first place. Those behaviors continued, and our Office of Science Director #2 said issues needed to be raised until resolved, that we "need to nip it in the bud."

I'd asked my Supervisor #2 on October 3rd, 2023, "What is the policy when branch leadership learns about problems with which they are involved?"

He said we lacked accountability, "I don't believe there is a branch policy for handling issues when they are raised. In the past, this was handled by the branch chief."

So, I said to the Associate Ombudsman, "We have federal employees who are pushing the people they supervise to lie in federal documents."

She said, "Yeah, I mean, obviously, captured in the SDR, it would never state that. It would be more so in the sense of like, 'This is how we did this in the past. This is what was affected. This is how it needs, like, the process needs to be done in the future. These are what the end results are.' Kind of something like that. That was obviously a very, very simplified version." She said the new Scientific Dispute Resolution process wouldn't include any mention of Executive Branch employees pushing those they supervise to lie in scientific regulatory documents.

She continued, "It's like, all of the, like, *flourishments* that you're saying, like, anyone who would read that at a leadership level, would be like, I mean, just so you know, also, too, even in an informal SDR, like, as long as you're okay with it,

CTP Director #2 and the CTP Deputy Director, they are made aware. Like, at this point, they're out of the loop of it all. But, yeah, even informally, if you were okay with it, the document would be shared with them too."

What? I had no qualms with sharing anything with leadership. Our training indicated we be transparent and raise issues. My job was to implement sound approaches and raise issues using clear language, so what did she mean?

Flourishment means "the act or state of flourishing" and to *flourish* means:

1. *To thrive or grow well.*
2. *To prosper or fare well.*
3. *To be in a period of greatest influence.*
4. *To develop; to make thrive; to expand.*
5. *To make bold, sweeping movements with.*
6. *To make bold and sweeping, fanciful, or wanton movements, by way of ornament, parade, bravado, etc.; to play with fantastic and irregular motion.*
7. *To use florid language; to indulge in rhetorical figures and lofty expressions.*
8. *To make ornamental strokes with the pen; to write graceful, decorative figures.*
9. *To adorn with beautiful figures or rhetoric; to ornament with anything showy; to embellish.*
10. *To execute an irregular or fanciful strain of music, by way of ornament or prelude.*
11. *To boast; to vaunt; to brag.*
12. *To brandish (a weapon).*

I'm pretty sure, considering she knew nothing about me, that she wasn't referring to 1 or 2. Number 3 was fun to consider, but didn't seem applicable. Number 4 didn't fit. Numbers 5, 6, 8, 9, 10, and 12 involved action she wouldn't know about. I could enthusiastically scribe and joust in my condo while working from home, but I wasn't on camera, so she wouldn't see that. Number 11? What was I boasting, vaunting, or bragging about? No, number 11 wasn't a fit either.

Did she mean number 7, "To use florid language; to indulge in rhetorical figures and lofty expressions"? *Rhetorical* didn't fit. *Lofty*, meaning tall, idealistic, or haughty? *Florid* means rosy or red in color, ornate, flowery, "in the bloom of health," or "in a blatant, vivid, or highly disorganized state," synonym: *psychosis*? Holy smokes, was I experiencing psychosis?!

What *flourishments*? I was using clear language. *To lie* is "to give false information intentionally with intent to deceive."

When I asked why concerns at CTP were ignored, she said she didn't know and that "There's just like, a very like, Wizard of Oz like man behind the curtain kind of like world that goes on."

8 | "In favor of industry."

I'D WAITED ABOUT A MONTH AND A HALF, until October 19th, 2023. Continuing to follow the Scientific Dispute Resolution process, the CTP Ombudsman and Associate Ombudsman then assisted in raising concerns to our CTP Director #2 in an SDR Initiation Memorandum. He was provided the same information about those nine EX REQ pathway assignments.

The four of us met on November 14th, 2023, and he seemed to understand the issues. I subsequently emailed him additional written evidence proving primary reviewers had been inappropriately blamed by leadership who directed staff to use deceptive arguments in regulatory documents. He wrote to me on December 16th, 2023, before he shared his decision, "I'm in the process of finalizing my determination with regard to the SDR memo, and it will be issued imminently. I do believe that the recommendations outlined in that determination will help address process considerations related to this matter."

Great!

However, his December 22nd, 2023, SDR Determination hadn't been helpful or truthful, and his recommendations and directives were never realized. Like the deceptive approach described in the SDR Initiation Memorandum, in his SDR Determination our CTP Director #2 had developed another false narrative. He used that to support his determination that the previously confirmed misrepresentation of data was, as he wrote, "a valid approach."

What?

Wow!

Like our assessments, he filled his inaccurate six-page SDR Determination with repeats of misleading, thin arguments that didn't support his determination. Instead, it seemed he could just *say*, with no real justification, that the FDA's use of fuzzy math was valid.

He wrote:

I have found that the use of cumulative marketing estimates approach is applicable and is appropriately documented in a footnote in programmatic environmental assessments.

Our Office of Science Director #2 had already indicated that deceptive approach wasn't anywhere close to valid. However, in his SDR Determination, our CTP Director #2 failed to mention that. He claimed in that determination that he hadn't seen evidence when he'd responded to emails directed to him to provide him evidence, and more information had been offered.

He wrote:

Further, Dr. Leppanen states that in instances when applicants have provided cumulative (or bundled) marketing estimate information, "Some branch leadership and some primary reviewers direct other reviewers to falsely represent marketing estimates as informative to an assessment of no impact in the assessment, intentionally devising a means to hide a confusing and additionally inaccurate explanatory statement in a footnote in a confidential appendix." As described above, the use of cumulative marketing estimates is appropriately documented in a footnote. I have not seen evidence that leadership has provided direction to reviewers to falsely represent market estimates or any other information.

He wrote:

I recognize that the bundled marketing estimates is a cumulative accounting of the marketing estimates of all products in the submission and not broken out by individual product. Moreover, when the submission is broken into bundles the marketing estimate information is not limited to that bundle. However, there is a footnote to the marketing estimates table of the programmatic assessments that clearly states what the marketing estimate includes (e.g., "Marketing estimates include the original product and all new products with the modifications listed in EX####### that the applicant intends to market under one name."). This information is appropriately contained in a confidential appendix of the programmatic assessment as applicant-provided market estimates information is commercially confidential information (CCI). While the information is in a confidential appendix, it is not confidential to reviewers or the applicant.

No, the footnote doesn't clearly indicate anything. And marketing estimates the footnote references are redacted and product STNs are truncated. That's part of the deception. More importantly, there is no indication anywhere in the entire assessment, redacted or not—confidential or not—that data combined for 32 products was never parsed to evaluate each of the 16 products purportedly evaluated. If that approach was up to snuff, why wasn't that explicit?

And our assessments weren't there to inform reviewers or the applicant. Our assessments were to inform decision makers and the public.

He'd even failed to mention that a solution had already been implemented, so the previous deceptive approach was no longer needed.

Further, he wrote:

As Dr. Leppanen acknowledged in our meeting on November 14th, 2023, the current practice of bundled application review and accepting cumulative marketing estimate information of a EX or SE submission has not resulted in any marketing authorizations that otherwise would not have been granted. As such, I find there is no adverse public health impact to the current practices the Office of Science uses in assessing marketing estimates to inform environmental impact assessment of product marketing authorizations.

What?
I hadn't acknowledged that.
How'd I even know that?

After introductions during that November 14th, 2023, meeting, our CTP Director #2 asked me to summarize what had happened. He said, "Okay, so I think what would be helpful for me, I kind of reviewed the documents, but if you could kind of give me like the *CliffsNotes*, a summary, and then I've got just a few clarifying questions because I want to make sure I've got all the information and that I'm not, you know, misinterpreting everything, you know. I'm mindful of time here, and we want to make sure we get a decision on this by December. I'm, of course, a scientist, so I'm really acutely aware of the science and so again I want the details. Defer to you on, you know, a higher-level summary and then I've just got a few questions that would be helpful for me to just kind of better understand."

In hindsight, his mention of "misrepresenting everything" made me laugh. But this isn't funny.

I said, "Okay, sure. And I'll do my best. It is a lot, but I will do my best to stick to the issue and be succinct."

He'd been at CTP for just over a year and I didn't know if he was familiar with our purview. So, I started with background, "The Environmental Science Branch is where I am, and we are charged with making sure CTP's National

Environmental Act Policy (NEPA) compliance is met. As part of that, we review applicant information associated with the environmental assessment, we do a review of information. An environmental assessment needs to be done for NEPA compliance. And according to that process, we look over information that the tobacco company provided, decide if we need additional information, and make a decision about—we're focusing on granting the marketing order and how that action, that federal action would, considering manufacture, use, and disposal—we evaluate how that action might affect, might have environmental impacts, to the human environment and the environment in general."

I then described how we considered marketing estimates. "As part of that process, we frequently consider marketing estimates. The applicant will provide projections of future market use, so five years into the future, one year into the future, and then also current marketing estimates. We use that information to make an evaluation of if there's an impact. So, for example, if an applicant is going to produce significantly more product, that might indicate or suggest there could be more, a greater potential environmental impact from manufacturing, use, or disposal. So, that information is important for us. We look at other things, but that's part of it."

I continued with a description of the deceptive approach, "In May, I received an assignment, with eight other reviewers, for a [tobacco company] bundle. We all had a similar application from, a similar set of applications from [tobacco company] and we were asked to work on this together. As part of that, our leadership and some staff advised us to misrepresent the marketing estimates that the applicant provided as informative to that assessment process, to the environmental assessment."

I was mistaken. I'd received the assignment in April.

I then described how our assessments were routed, "We prepare that and then that assessment is forwarded to our division leadership, and [our Division Associate Director] would sign—review the assessment and make a decision about if there'll be findings of no significant impact—and he would sign that if that was the case. If there were findings of a potential impact, we would then move to an environmental impact statement. So, we put this information about marketing estimates into that assessment and forward that to [our Division Associate Director] for his evaluation. In this case, there were a number of instances where the, what the applicant provided, wasn't helpful to our evaluation. And there were two team supervisors or two acting branch chiefs, and one team supervisor or one other reviewer [referred to as such because their positions changed], and several others that directed the rest of us, when we were protesting, to use this information as if it represented accurate information."

I reiterated the simple story, "In short, we had marketing information for 32 products, and they wanted us to present it as if it informed 16 products."

I discussed concerns, "This caused a lot of problems. Some of us did it. Some of us didn't. I've tried to raise this issue for over about six months now. Through that process, I didn't know at the time, but I learned that this had been done previously. The primary concern here is about our regulatory compliance, our assessments, how they're prepared, and if they could withstand scrutiny. If there was any sort of litigation, these documents are made for public, to inform the public about our decisions. That's what the NEPA process is for. I think one concern here is if those documents were evaluated, I don't believe that those that took this approach would withstand scrutiny. Another issue is that staff were asked to do something that a variety of staff indicated was, we considered to be unethical."

Then I paused, "I have a couple of ideas as far as solutions, but maybe I'll stop there and see if you have any questions."

He said, "Okay, yeah extremely helpful. So, appreciate that primer. Just a few clarifying questions just so I can better understand. So, what is the rationale for aggregating them? So, was that just the volume that was coming through, you know, what is the explanation for the aggregation of, across all of the applications?" He was asking why marketing estimates were aggregated across all the applications, but they weren't. Only some were aggregated in groups of 32. I didn't know why. Regardless, branch leadership had asked the nine of us to agree on a consistent approach across those nine EX REQ pathway assessments.

I said, "The nine of us were working independently, but were asked to come to a consensus on how we were going to do it. So, the *applications* weren't aggregated, per se. It was nine different assignments, but because we all had similar—because the applications were *similar*—we were asked to come up with a similar approach and to be in agreement on that."

"Okay, and it sounds like there's other people that were not in agreement with this approach?" he asked.

I replied, "Yes, there were several of us. I did not end up using the marketing estimates and one other reviewer did not end up using the marketing estimates at all. So, we excluded it from our assessments. Other reviewers had different takes on it. Some felt pressured. And some of them hadn't passed their one-year probationary period. Some, one other person said he was nervous about raising the issue because he was working with a temporary supervisor and he didn't want to make things difficult for her. We had a lot of discussions about this over the course of about two weeks and that, a lot of that, some of that has been captured in Microsoft Teams chats, and in emails from our leadership directing us to do this in instances where we were indicating that this was a false representation of the data."

"Okay, and so in terms of the footnote, that basically has just been acknowledging the approach that has been taken, and there is potential for bias?" He further asked, "Or what's the extent of the footnote that's presently being used on this issue?" He referred to those deceptive footnotes that obfuscated marketing estimates combined for 32 new products, where half of those products weren't part of the assessments where the footnotes were used.

I explained, "The footnote varies between different reviewers. So, it can be a little bit different. It was different within nine reviewers working on a variety of bundles. And then when I looked back into the record and I found previous practices of this, it was a little bit different. But the footnote, I think, some have indicated they think that it releases them from any liability when it comes to the fact that they did something. It's meant to describe, I think, that the data don't represent the data in the application, but that's not clear."

It was a sneaky way of indicating the data didn't apply, otherwise it probably wouldn't have been referenced in a Microsoft Teams chat as a means to "cover our butts." Some argued that the deceptive approach was legitimate because marketing estimates associated with each of the 16 new products to be considered were part of estimates for 32 new products, and that the footnote provided an explanation about associated uncertainty. But it didn't.

If the footnote meant that, why not just say that? Why not simply say, "The agency's conclusions are based on an assessment of data combined for 32 products. There is some uncertainty associated with this assessment because the agency's determination applies to only 16 of those 32 products"?

Further, hypothetically imagine the future marketing estimate combined for the 32 new products was 34,000,000 units. How would a reviewer decide to split 34,000,000 units among 32 products to get at data applicable to each of the 16 products they were to consider? By guessing?

And this pointed to a larger issue. We were to use credible science that included appropriate representations of uncertainty, but we weren't doing that. So, I said, "And I think one concern that we had is, and we've talked about this, and I don't want to go off subject too much. But if we have uncertainty, should we include that in the assessment clearly, explicitly?"

He was a scientist and talked a lot about being a scientist, so he should've easily understood that. And this broached on solutions we could implement. If we'd been honest in the footnote, we could've said, "Data combined for 32 new products was used to support the agency's determinations about 16 different new products." But if we'd been forthright, as our oath and agency training required, the footnote might've said something like, "Data combined for 32 new products are included but cannot inform the agency's determinations about 16 different new products." And if we'd *really* wanted to follow our oath and agency training, we might've included that in larger text, and maybe not in a footnote in a confidential appendix.

"Instead, and I think if you look at the practice and when this happens," I reiterated, "it's not clear. And so, the footnote. I think, reviewers thought, 'Oh, this gets me off the hook for what I'm doing.' I think that's what the intent of the footnote was. And again, this is captured in Microsoft Teams chats."

We were on video and he was writing and said, "Okay, got it, and I'm taking notes. I'm not multi-tasking here."

He continued, "So, can you tell me a little bit about, so if we had done things differently, do you think that there would have been a change in any of the determinations for any of the applications? I'm just trying to understand, like, the severity of what we're potentially looking at. There would be a different determination as a result of us doing something differently with regard to how we handle these?"

His mention of "a different determination" referred to the agency's findings. If any of those determinations had been done differently, those marketing orders might not have been granted or might've been delayed.

I said, "I don't think so," in reference to several of those nine EX REQ assignments I was more familiar with. And I immediately followed with, "So, I haven't reviewed the record completely, right? But to my understanding, is that if we could have just excluded it [the deceptive approach], then we could have produced an accurate document with the same outcome."

Excluding the deceptive approach would've excluded deceptive elements, and therefore each assessment would've been accurate and consistent, but not necessarily scientifically sound. Our Division Associate Director had reviewed those nine assessments, some that included and some that excluded the deceptive approach, yet he made the same determination in each case. I wasn't familiar with how he reached those determinations, but I knew the outcome was the same. My point was that we could've excluded the deceptive approach and reached the same outcome, because that had happened in those two cases where the deceptive approach had been excluded. We hadn't needed to use the deceptive argument. Our Division Associate Director made the same determination without it.

He further asked, "Ok, got it, got it, got it. And so then in terms of, I'm just thinking through like the directionality of any bias. So, would the current approach be a *conservative,* or would it be *overselling,* or would we not know, like talk me through, like, *directionality* of bias as a result of what we're doing in terms of determinations on the applications."

I replied, "Sure, those applications where this happens, it always, the information that's presented, is always in favor of a marketing granted order. And it is misinterpreting the information, one might say, on *behalf* of the tobacco companies. It never goes in the other direction. So, that's one thing that we're concerned about as well."

He reiterated, "Okay, so it would be conservative *in favor* of industry, of the applicant, then, I guess."

I said, "Yes."

He replied, "Okay, that's helpful. Good, good, good. Okay, so, I'm interested in your thoughts around, like, correcting the record, you know, as you noted in the, you know, memo and it seemed like at the onset you said you kind of had some ideas. So, I'm just interested, just want to pick your brain on, you know, what you feel would be the ideal scenario here in terms of retrospective, but like also moving forward. So, interested in your thoughts on that front."

"Sure. Yeah. I think, I've been thinking about this. And I think there are three things that off the top of my head need to be considered." I said, "I think one thing, we need to understand the breadth of the problem. So, as I mentioned, I had looked back into the record a bit and I noticed that it happened in the past, I didn't know this initially. So, I think in order to understand how we correct the record, we need to understand what it is. And that could be done. I did a good tutorial yesterday on the new CTP search tool, and that's how I found some of this in this first place. I don't think that would be an impossible task to get a feel for that. If it is as it appears to be now, I think it's possible that there could be some, and this again, this is not my expertise, but maybe there could be something put into the record acknowledging that this mistake was made or however it would need to be posed, that it had been acknowledged. But the issue would be is, if we did find cases where the approach, without having that information [data parsed by product], would change the decision. And I think that's what we need, we need to look for cases where that would change the decision. Again, I don't think that's going to be likely, but I cannot speak to that completely. So, I would think something like that as far as how to correct the record, and again, a number of us are concerned about that record cause it's discoverable. And these documents, again, are for public consumption to inform the public. So, that's, there's an additional caveat there."

It certainly didn't feel like we were meeting our obligation.

I continued, "The other two things, the other two are specific training or guidance and more general guidance. And so, I think for our division and branch, we need specific guidance related to use of marketing estimates that indicates how it can and cannot be used. And then general guidance is important from the perspective that all staff, to my understanding, all staff shouldn't be pressured to do anything that they consider to be unethical or inaccurate or misleading." Our CTP Office of Management Deputy Director, who later became the Director #2 of that office, told me this when she saw evidence of the deceptive approach and other concerning leadership behavior.

"So, that," I said, "those three things are the things that I've focused on so far." Those three things were: 1) Implement guidance about how we use marketing estimates, 2) Implement guidance about scientific integrity, because scientists were pressured to lie, and 3) Do a proper review of the record to inform any needed corrections.

He further asked, "Okay, that's helpful. And is there anything that can be done in terms of more precision, you know, moving forward so that we're not, you know, aggregating marketing estimates across applications and that we, you know, can make sure that we're dealing with, you know, product-specific information when making determinations? You know, what is the ability for us to modify existing protocols? So, you know, the training is good, the guidance, but I'm still trying to, you know, better understand, like, what is it? What are we telling the reviewers to do? And is there a change in our SOPs to accomplish that, if that makes sense?"

I said, "The ability for each individual reviewer to raise their concerns. I think if that happened and we had that collective body of information, then through that consensus among the scientists that are doing these reviews, it would prevent this sort of thing from happening again. So, this issue was somewhat specific, but if we think about what information we need to prevent something in the future—I think it's just, in this situation, we had two of us [who] were very vocal. There were others that were vocal in less pronounced ways—if we created a platform, something in our branch, where we made sure the concerns of all reviewers were just documented in some way to make sure we weren't stifling certain scientific opinions and we didn't—there was a very top down approach to how we did this—and so—"

He asked, "And for the—sorry to interrupt—that's top down from the division or from the office level or both? Just so I can understand that scope."

"That was from the branch level," and I continued, "I think there are other issues. You're asking about bundling and applications, and there are a number of different issues as far as that goes. When we receive these assignments, right away we know what limitations there are." We did not talk further about that then, but emailed later about limitations, for instance, related to bias, uncertainty, and data gaps and other excluded information.

I continued, "In this case, it's not anything necessarily that the applicant needed to do differently, and," referencing how CTP bundled applications, "it wasn't even anything necessarily that CTP needed to do differently. It was just the ability for the very valid concerns of the reviewers to be heard. I think that is the best solution that will get at, again, this is a specific instance, but that would prevent any sort of thing like this from happening in the future. Because our branch is growing and we've got some good people that will speak up with evidence and want to talk about things."

"Okay, no that's helpful. So, it sounds like a more *process-oriented* resolution or at least what I'm hearing from you is just making sure we've got mechanisms in place for people to be heard, to document that would be a helpful, at least headed in the right direction. Is that right?" he asked.

I answered, "Yes. Yep, I think so."

I hadn't "acknowledged in our meeting on November 14th, 2023, the current practice of bundled application review and accepting cumulative marketing estimates of a EX or SE submission has not resulted in any marketing authorizations that otherwise would not have been granted."

There was no way I'd know that. That would require we review our work products and associated materials submitted by each applicant for all potentially relevant EX REQ and SE pathway applications. I'd been clear that "we need to look for cases where that [correcting the deceptive approach] would change the decision." This would number into the thousands, especially if we also considered SE pathway applications.

And why had he added mention of the SE pathway in his determination?

I'd only referenced our use of the deceptive approach in our reviews of applications submitted via the EX REQ pathway. Our meeting involved only EX REQ pathway applications, and no mention was made of the SE pathway. And in his SDR Determination, our CTP Director #2 had even written, "The information and examples Dr. Leppanen provided were for EX applications."

Holy smokes! I looked into the record again and the deceptive approach had been used for SE applications, as well!

This looked like damage control.

Interestingly, despite the deceptive approach he used to feign a credible SDR Determination, he also wrote in that determination that there would be improvements. But if the approach had been valid, why would we bother? Why would we dedicate efforts to unneeded changes while leadership complained about our demanding workload and need for additional hires?

For example, after he indicated the deceptive approach was valid, he wrote:

However, as discussed below, I have determined that there is an opportunity for greater internal clarity on how to treat marketing estimates in various circumstances. Therefore, I am directing the Office of Science to take steps to update job aids and review guides, as appropriate and in consultation with the Office of the Chief Counsel, to provide greater clarity on how marketing estimates should be treated in various scenarios.

After he said fuzzy math was valid, he wrote that many times, for instance:

II. Response to specific requests made by Dr. Leppanen

In Dr. Leppanen's memo, she makes specific requests. Below, each request is summarized followed by a response and explanation for the response.

1) Bundled application review in circumstances when applicants provided cumulative marketing estimates result in flawed programmatic environmental assessments. The record should therefore be corrected where this practice has been used.

Response: As discussed above, I have found the current approach is valid and cumulative marketing estimates appropriately documented. Thus, there is no need to correct the record.

2) To prevent further applications of this practice, guidance to prevent false representation of marketing estimates should be developed and implemented.

Response: As discussed above, cumulative marketing estimates have been appropriately documented in programmatic environmental assessment. However, consistent with my decision, the Office of Science will be updating reviewer resource materials/guides, as appropriate, to ensure consistency and provide clear direction on what marketing estimates are necessary for an adequate assessment in various circumstances.

He wrote again:

Conclusion:
As Dr. Leppanen acknowledged in our meeting on November 14, 2023, the current practice of bundled application review and accepting cumulative marketing estimates of a EX or SE submission has not resulted in any marketing authorizations that otherwise would not have been granted. As such, I find there is no adverse public health impact to the current practices the Office of Science uses in assessing marketing estimates to inform environmental impact assessment of product marketing authorizations.

However, I do acknowledge that there are opportunities to provide additional clarity for how to treat marketing estimates in various circumstances. For example, for which pathways are product specific marketing estimates necessary for an adequate assessment, when are cumulative marketing estimates sufficient for an adequate assessment, and how to document it as such (e.g., appropriate footnotes). Therefore, I am directing the Office of Science to update relevant reviewer resources to provide additional clarity on how to treat marketing estimates in various circumstances and as appropriate consult with the Office of the Chief Counsel. In parallel with these efforts, assessment review, programmatic assessment development, and environmental impact determinations for product review should continue.

He promised improvements because he, like our Office of Science Director #2, knew the approach wasn't valid. If he hadn't said the deceptive approach was valid, our review and assessment process would go on hold while past and current practices were investigated. And without our assessments, marketing orders couldn't be granted.

So, the outcome of the first implementation of CTP's new Scientific Dispute Resolution process confirmed CTP's ethos under our new CTP Director #2. This had been an opportunity to realign our work. Instead, our new CTP Director #2 agreed that we could concoct and use deceptive arguments in our regulatory science documents, and that CTP leadership would do the same in its internal agency communications. It confirmed that leadership could, and would, lie.

Now, our CTP Director #2, as well as the CTP Ombudsman and Associate Ombudsman who also attended that November 14th, 2023, meeting, knew that I knew that our CTP Director #2 was lying. They and all other informed leadership knew we were up to no good. Our CTP Deputy Director, who was documented as part of discussions associated with the inaccurate SDR Determination, had also been made aware.

My oath and training still indicated we should do things differently.

9 | "No definitions or duties"

AFTER SEVEN MONTHS OF STONEWALLING, on December 14th, 2023, Supervisor #2 had finally engaged about leadership's directives to lie. He claimed, "It's up to the reviewer to decide the approach they want to take."

That was ludicrous. Primary reviewers had no means to independently decide what was included in regulatory reviews or assessments or in any other work products. We simply didn't have any sort of autonomy. Were we supposed to? Nonetheless, Supervisor #2 was blaming primary reviewers who used the deceptive approach, despite written directives from our leadership, our secondary and tertiary reviewers, to use it.

And if it was a "valid" approach, why the hubbub—why point fingers—and why had an alternative approach been developed and implemented? If it was a valid approach, why were primary reviewers being blamed for doing something wrong? I hadn't used it because he'd offered the option to leave it out. Why leave it out if it was "valid"?

All this deception had created a mess. Fuzzy math couldn't be used again, but they had to pretend it was valid to avoid admitting it happened and correcting the record. However, because it was invalid, those who directed staff to use it tried to find someone else to blame.

"That guidance needs to be in writing," I said. "That is not an understanding among the primary reviewers in our branch." I'd never seen that policy. He claimed that primary reviewers had independently decided to use fuzzy math and deceptive arguments in FDA regulatory decision documents about tobacco products. This was a big deal, so I said, "That's a big deal." I said, "If it's actually up to the reviewer, that needs to be in writing."

He responded with a noncommittal, "Okay."

I continued, "Because the understanding is that the secondaries are reviewing and making sure things are cohesive across the branch."

"Right," he said.

"What's needed in our branch is guidance," I reiterated. "I just think it's a little bit of passing the buck. 'The primary reviewer did it. I didn't know.'" I said, "You guys are secondaries, you've got a branch chief. It goes through a process."

This was confusing. He said it was "up to the reviewer" but there was a process in place that designated leadership guided the content and outcomes of those assessments. To what extent they *should* was another long-standing question. However, the role that secondary and tertiary reviewers had to control and limit the scientific contributions of primary reviewers in work products signed by primary reviewers was clear and unequivocal.

I said, "This needs to be dealt with now, in writing."

"I think that's going to be difficult," he said, "for a variety of reasons. I mean, and I'm talking about administrative-type reasons not scientific-type reasons, because I know we've tried to talk, we've talked about having guidance in the past and there are some issues where those things become FOIA'able." He was referring to the Freedom of Information Act or FOIA, where the public has access to "FOIA'able" information.

He explained, "And so we were, we were, well, we weren't able to do it. It doesn't mean that it won't happen. Unless something comes down from [our CTP Director #2], you know, where we have to put something together, but it's gonna have to be done very carefully."

I increasingly noticed specific issues were often framed in larger, daunting contexts therefore too big to handle, and this didn't need to fall victim to those tactics. So, I refocused our discussion, "The only thing I'm talking about [right now], 'What is the role of the primary reviewer?' Fill in the blank. 'What is the role of the secondary reviewer?' And you just said it to me."

He'd just said what was in the assessment was up to the primary reviewer, so why not put that in writing? Instead, we were supposed to use word of mouth or just *know* about what was a consequential unwritten policy? Reflecting in writing what happened would look bad: "The outcome of your assessment shall support the agency's granting of marketing orders for the new products. Therefore, select for inclusion in your assessment information that supports the agency's granting of marketing orders for the new products."

He responded with a flat "Um hmm" that I couldn't characterize as agreement or disagreement.

"So, we can't have in writing, 'What is the role of the primary reviewer?' We can't have that? Do other branches have that?"

He said, "I don't know."

"We literally cannot have in writing, 'What is the role?' You're basically saying that it's the primary reviewer [that decides] and the primary reviewer gets blamed but there's no policy. Anyway, this is nuts. This is absolutely nuts. If higher up at CTP they knew about these issues, and the fact that it's, I mean—How long has the branch been in existence?"

He said, "I guess 12 years."

"Twelve years," I said, "and there's nothing in writing about what a primary and secondary reviewer do? Maybe there's something out there that we can pull up. I'm guessing there must be, right?" If we'd been doing this for 12 years, it had to say somewhere that the primary reviewer's role was to determine the approach used in our assessments.

I continued, "We must have something in our branch SharePoint [shared files], just to look at that and say, 'Ok, is this what we do in practice and does it need to be updated?'"

He said, "I'm going to reach out to [our Branch Chief #3 and #5] to see if she knows if there's something like this. So, I'll see what I can come up with."

The next day he wrote, "To answer your question that you brought up in our one-on-one meeting, after speaking with [Branch Chief #3 and #5] and [Branch Chief #4], there are no definitions or duties specified for primary and secondary reviewers."

During our call the day before, I couldn't shake the feeling that we had something. If it was really "up to the reviewer to decide the approach they want to take," why had we received those directives to use the deceptive approach? Now he'd written that "there are no definitions or duties specified for primary and secondary reviewers."

So, I asked a colleague. Of course! The document he sent was frequently referenced by all leadership and described specific duties of tertiary, secondary, and primary reviewers in all scientific disciplines:

Tertiary Reviewer

- *Senior-level discipline scientist (branch chief or above)*
- *Reviews discipline review decisions and conclusions, asks questions and has a discussion with the primary reviewer to ensure the conclusions are consistent with the discipline's current thinking*
- *Reviews conclusions for scientific accuracy, adherence to division/office policies, and consistency in how scientific evidence is weighed within the discipline and across the office*
- *Provides broader review perspective of the deficiencies from the discipline to ensure adequate strength of scientific findings*

- *Provides senior-level concurrence on the conclusion and findings of the primary reviewer*
- *Provides feedback within the review timeframe via a meeting or written comments*

Secondary Reviewer

- *May assign primary reviewer*
- *Participates in recurring coordination meetings with the primary reviewer to ensure discipline issue resolution and escalation*
- *Conducts secondary review of FDA work products, as needed, including:*
    - *Provides technical feedback on the review*
    - *Ensures the review aligns with the discipline's existing guidelines, scientific consensus, or precedent, as appropriate*
    - *Provides stylistic or editorial feedback only if needed to improve clarity*
    - *Ensures information in the review, such as product information Submission Tracking Number (STN) references, are accurately captured*
    - *Flags issues for the tertiary discipline reviewer's focus (e.g., analysis of data that informs the basis for a deficiency)*
- *Assists primary reviewers in resolving issues that arise during scientific review*
- *Identifies potential regulatory and policy issues that may compromise timelines*
- *Tobacco Product Master File (TPMF)*
    - *Ensures the TPMF is reviewed within the scope of the application reference*
    - *Ensures the discipline application review does not contain confidential information that should be in the TPMF review and vice versa*
- *Signs the application and TPMF discipline reviews as the concurrence signatory*

Primary Reviewer

- *Conducts discipline-specific scientific application and TPMF reviews within established timelines of the application program*
- *Verifies product unique identification information for accuracy, and notifies Regulatory Health Project Manager (RHPM) and Technical Project Lead (TPL) of changes needed*
- *Initiates consultation requests needed to support review, including requesting Customer Service Center assistance with reviews involving data automation and analysis*
- *Works with primary reviewers from other disciplines, the RHPM, and the TPL*
- *Schedules small group meetings to ensure conflict resolution on potential overlapping deficiencies*
- *Identifies and raises unresolved issues to discipline leadership and the TPL*
- *Contributes to review-related briefing materials*
- *Drafts external meeting request responses*
- *Identifies issues or deficiencies in consultation with discipline leadership*

- *If applicable, application and TPMF primary reviewers work together to determine whether and how the TPMF(s) support(s) the referencing application*
- *Updates the TPMF tracker with review information*
- *Submits requests to the TPL for any timeline changes, and includes rationale to support the request*
- *Uses the current version and format of the discipline review template for drafting the review (it may be in Rhapsody* [CTP's regulatory review software application], *but always check the Reviewer Resources site version, which supersedes all versions found elsewhere)*
  - *Flags any template deviations for TPL consideration*
  - *Submits template and job aid suggestions, questions, corrections, or anonymous feedback on the Reviewer Resources site*
- *Ensures that any edits to their discipline's language in the Deficiency letter are updated appropriately in their discipline review*
- *Signs the review*
  - *The TPMF primary reviewer signs the TPMF discipline review*
  - *The application primary reviewer signs the application discipline review*
- *Uploads final discipline reviews into the appropriate database*
  - *EX, SE, and PMTA in Rhapsody*
  - *TPMF in ToRIS*
- *Ensures accuracy of uploaded reviews and provides corrections or addendums to respective reviews, if discrepancies are identified*

We certainly did have "duties specified for primary and secondary reviewers." To avoid their own culpability, leadership blamed employees who followed their directives as this guidance indicated we do. Supervisor #2 was lying. He did that despite written evidence that contradicted what he said and wrote.

My Supervisor #2 reviewed and signed documents that included the deceptive approach. One person he consulted about any guidance was our Branch Chief #5 who was also our former Branch Chief #3, our Team 2 Supervisor #1 and #3, and a longtime employee with historical knowledge about how we did things. When she was our Branch Chief #3, she oversaw earlier implementations of that deceptive approach. So, she reviewed and signed documents with the deceptive approach. And she'd subsequently written in support of later uses of the deceptive approach when we were directed to use it by the other person that my Supervisor #2 consulted. That other person, our Team 3 Supervisor #1 and #3, had been in the branch for over six years and was also our former Branch Chief #4. He'd written extensively about why we should use the deceptive approach when he directed us to use it. The collective knowledge of these three individuals in our branch totaled 25 years, yet they falsely claimed, "There are no definitions or duties specified for primary and secondary reviewers."

These individuals guided our approaches to ensure every single regulatory document that we produced followed "the discipline's current thinking."

No regulatory document moved ahead without reviews, edits, and approvals by several layers of leadership and administrative staff. Secondary and tertiary reviewers were part of our leadership, the people who also evaluated our performance. In most cases, my Supervisor #2 had also been my Secondary Reviewer #1. He and my other secondary reviewers advised me how to proceed. Our branch chief, who was usually our tertiary reviewer, also chimed in to direct us how to proceed.

They controlled that process.

That practice is documented via email chains and documents they reviewed, edited using "track changes," commented on in comment bubbles, and signed. A History and Concurrence Page was even attached to every single one of our reviews that documented the dates that each individual at each level reviewed, edited, and concurred with the document, back and forth through many iterations while the document was developed.

Individuals from our Science Policy Branch and Regulatory Health Project Managers and Technical Project Leads serving as quaternary reviewers chimed in as well. They and other leadership like our Division Associate Director, Deputy Director for Regulatory Management, and directors of other divisions, decided which concerns continued into or were excluded from those assessments that the public saw, but that included as a testament to their credibility only the names and credentials of primary and secondary reviewers. When those work products diverged from those credentials, CTP nonetheless still publicly claimed those credentials applied, so maybe reviewers were *supposed to* have some say.

Anyway, I emailed our Division Associate Director and copied the CTP Ombudsman, "I did not use the questionable approach, but I will not stand for my colleagues being blamed for approaches that branch management directed staff to use—as per guidance in their roles as secondary and tertiary reviewers, guidance that branch management is pretending does not exist, it looks like to cover themselves at the expense of others."

I requested our Division Associate Director to "Please request Alternative Dispute Resolution services for our branch." Such services were perfect for such scenarios, but staff at my level couldn't request services for the branch, only management could. However, he did not respond.

Because the SDR process was underway, I forwarded the same information to our CTP Director #2 and to our Office of Science Director #2. This was a continuation of the same behavior previously reported to them and they'd welcomed follow-up.

Supervisor #2 was moved to a non-supervisory position and no longer interacted with us. Team 1 was then consecutively assigned five different short-term acting supervisors during the remainder of 2024, some exhibiting more extreme questionable behaviors. The Team 3 Supervisor #1 and #3, who'd been our Branch Chief #4 and directed us to lie, left the FDA a few months later. Our Branch Chief #3 and #5 returned to her role as the Team 2 Supervisor #3. A reviewer she supervised, who was also complicit when she was in a temporary role as the Team 3 Supervisor #2, became our permanent Branch Chief #7. The reviewer who showed us how to "cover our butts" moved from Team 2 to become the permanent Team 3 Supervisor #5.

Deck chairs had been rearranged, and the bad behavior continued.

And what was our *current thinking*?

It was that our assessments, no matter what, always facilitated marketing orders. Millions of applications had been submitted for marketing orders for tobacco products. Of those, an assessment would've been completed for every product that received a marketing order. Yet no assessment in CTP's history had ever been associated with anything but a finding of no significant impact.

We cut and pasted, cherry-picked, and rubber-stamped. The questions we were allowed to ask of applicants and the elements that we included in our assessments were carefully selected every time to support findings of no significant impact. Considerations were excluded that might raise red flags or require a robust evaluation that would take more time. It was our *current thinking* that would be problematic for the public to discover via a FOIA request.

Our regulatory reviews that framed our subsequent assessments included a statement attesting to the accuracy of those efforts. That statement said, "I attest the contents of this review and evaluation are accurate to the best of my knowledge." Primary reviewers were provided the option to remove that statement. While it might be difficult to argue that not having a statement like that is fraudulent, having it and removing it to placate concerns might be.

10 | Figure It Out

OTHER ISSUES HAD BEEN RAISED about products that didn't receive a full, substantive scientific review, including those where the fuzzy math had been used. New products processed via SE and PMTA pathways were to undergo comprehensive scientific reviews involving disciplines such as Toxicology, Pharmacology, Microbiology, Epidemiology, Engineering, Social Science, Chemistry, and Environmental Science. However, EX REQ reviews typically only involved Chemistry and Environmental Science.

Reviews limited to only Chemistry and Environmental Science disciplines complicated work for my branch. The Environmental Science Branch, where I worked, was often not allowed to ask an applicant questions unless other disciplines involved in the review of the same products had questions. With only one other discipline involved, the likelihood we could piggyback our requests alongside another discipline's questions was greatly reduced. If no other discipline had a question, some leadership said we needed to "figure out" how to facilitate marketing orders for the new products without the information we needed to perform our assessments.

That said, CTP welcomed some questions from Environmental Science when the agency wanted to *deny* a marketing order, explicitly to help build the case for that denial—to stack the deck.

That created inconsistencies in what we could ask, and therefore in how our reviews and assessments were performed. I was able to ask few clarifying questions in my work on EX REQ pathway applications and the most via work on PMTA applications slated for denial. So, the same sorts of issues were handled differently across different applications. This created confusion when a reviewer would raise an issue because that had just been allowed in one review, but then it wasn't allowed. And because our work wasn't consistent, we had no precedent. Assessments instead built cases for predetermined outcomes via various means.

Exemption from substantial equivalence (EX REQ) was considered our most extreme form of rubber-stamping. It provided an expedited application process where some products were exempt from many considerations the agency had determined were important for the protection of the public health.

Those expedited reviews helped the agency meet its metrics when Congress wanted to see that applications had been processed. However, it didn't ensure the agency's public health mission had been met. In some cases, new products differed from their comparator products in ways that, scientists had indicated, required a full scientific review. The agency granted hundreds of marketing orders under the expedited EX REQ pathway when we needed to evaluate at least some of those products under the more substantive SE pathway.

By early 2025, some leadership was finally paying attention. CTP was grappling with a means to change gears moving forward, to perform substantive reviews in cases like that. However, what would come of products already granted marketing orders that lacked those needed substantive reviews and had been sold for years?

Leadership said we couldn't consider how changes in what we did, or even new science, might apply to products already legally marketed. That would mean newer products would be delayed by substantive reviews, while already marketed products not subjected to the same scrutiny would remain on the market. So, nicotine products for sale in the U.S. are a hodgepodge of illicit products about which we know nothing alongside products that underwent various means of regulatory review.

And our leadership misinformed us. CTP *could* reconsider products that were already legally marketed.

11 | No *New* Impacts

OTHER CONCERNS HAD BEEN RAISED about how CTP handled assessments of products already sold in the U.S., when products were sold without marketing orders. CTP hadn't yet identified or evaluated exposures or impacts associated with the manufacture, use, or disposal, of those products.

In those cases, some assessments written by my colleagues concluded there would be no *new* impacts because the products were already being sold. When I raised this concern to my Supervisor #2, he said I could exclude the argument from *my* assessments, but some other reviewers continued using it. Those assessments would say something like, "The agency determined there are no new impacts because the new product is already marketed in the United States."

No duh.

Of course there'd be no *new* impacts from granting a marketing order to sell a product if the product was already sold. The point of that comparison and the review process, in general, was to complete the review and assessment *before* the product went on the market. That way people would be informed ahead of time.

And how far could this sort of rationale extend in the federal government?

How about building a pipeline that leaks *before* doing the assessment, so when the assessment is done the agency makes a determination of no *new* impacts, like CTP did, because the leak happened before the assessment. Imagine what that might look like, "The agency determined there are no new impacts associated with the decision to construct the pipeline because it has already been built and leaked and killed a bunch of fish and babies that drank the groundwater that it contaminated. Because the fish and the babies died *before* the assessment was completed, there are no *new* impacts requiring consideration." Approved!

Sorry babies!

Brilliant! Applicants could simply sell their products before marketing orders were requested. That would ensure a determination of no *new* impacts and CTP's work was made easier. CTP staff that came up with these arguments acted like they were special geniuses. Kudos!

If the review and assessment happened *after* a new product went on the market, it might've made better sense to consider outcomes resulting from a marketing order *denial*, to consider what might happen if the product would be removed from the market. This could result in some benefit we never considered. And imagine a marketing estimate of 2,000,000,000 cigarettes or 2,000,000,000 ENDS products with impacts that weren't considered because marketing orders came after the products were on the market.

In his December 22nd, 2023, SDR Determination, our CTP Director #2 wrote that our assessment was "a process for considering the impact on the human environment of a proposed agency action and an alternative (i.e., authorizing the marketing of a new tobacco product vs. not authorizing its marketing)." So, in our assessments, we were supposed to compare scenarios associated with the manufacture, use, and disposal of each product when it was and wasn't sold in the U.S. This is the information the agency had determined it needed to inform its decision and the American people.

However, we weren't doing that, and we weren't considering the alternatives. Some assessments would just say, "We assume no impacts," and not bother considering in any way what not authorizing a product might mean.

12 | "No threshold is possible now or near future."

ANOTHER CONCERN WAS THE AGENCY'S POLICY to exclude considerations of thresholds in our regulatory science work. In this context, a *threshold* is a level above which an effect is likely. A *threshold* is defined as "a level, point, or value above which something is true or will take place and below which it is not or will not."

For example, if the threshold beyond which your consumption of peaches makes you sick is 23 peaches, and you eat more than 23 peaches, then it is likely that you will get sick. Such thresholds are important because they help communicate how one might prevent a possible negative outcome of an exposure. Considering our peach example, the exposure is your peach-eating. You could likely avoid getting sick by eating less than 23 peaches.

In regulatory science, thresholds support sound scientific decision making and disclosures. If a public health agency knows you will likely get sick if you eat more than 23 peaches, they can tell you that. They can alert you to possible outcomes. You could then moderate your peach consumption in a way that means you'd likely never need to get sick from eating peaches. If you decided to eat more than 23 peaches, you could anticipate the outcomes. Your peach-eating responsibility shifts to you, and so does your power and freedom to make informed choices. Although overeating peaches might benefit peach growers, the agency's purview is to protect you, not the peach industry.

CTP's disregard of thresholds came to light when staff asked how arguments used in our assessments were interpreted when the agency determined findings. Our assessments almost always pointed to some percentage of the overall market that the new product made up as supporting the agency's claim there were no impacts that we or the American people needed to worry about. Assessments would say something like, "The new product constitutes only 3% of the existing market, so the agency has determined there are no significant impacts." In the same context, assessments would sometimes also refer to that percentage as "minuscule."

However, we didn't consider in any way what percentage might be associated with any sort of effect. We didn't describe or consider what was meant by *minuscule*. Therefore, there was no scientific basis behind those claims.

And the market was changing. In one case, a new product constituted more than 10% of the existing market. Was that "minuscule"? It seemed plausible that might affect something, but we didn't even consider that. And what about new product categories where the first ever product marketed therefore constituted 100% of the market? Would the agency say, "The new product constitutes only 100% of the existing market, so the agency has determined there are no significant impacts"?

If marketing a new product might contribute to crossing a threshold from no likely effect to one that might happen, we'd need to consider that. We might need to communicate that possibility to decision makers and the public.

Considering our peach example, it was like we hadn't bothered looking into how many peaches would make people sick. Instead, in the assessment we'd say the agency found no significant impact associated with people eating peaches and point to a confidential appendix. However, the number of peaches was redacted in that confidential appendix. Additionally, it was never even considered if that undisclosed number of peaches would make people sick.

In some cases, we didn't need to determine those thresholds ourselves. For a suite of contaminants, we could use thresholds already developed and vetted, that were frequently used when risks were assessed and communicated. That is common and a very normal thing to do in these very sorts of circumstances. It is fair to say doing that in this sort of scenario, especially as a public health agency regulating tobacco products, is expected.

Considering our peach example, this was akin to the U.S. federal agency that regulated peach consumption, for the benefit of the public health, excluding any mention or even any consideration of the 23-peach threshold, that they know about, from their assessment meant to inform decision makers and educate peach eaters about what they could do to avoid getting sick from eating too many peaches.

On November 14th, 2023, our Division Associate Director wrote that "Thresholds for marketing estimates are not required to develop the assessment. To remain consistent with our regulatory work, CTP's assessments have been written for many years (not expecting you to know your branch history) with marketing estimate projections without use of a threshold. This is our current practice."

He wrote, "A threshold would be arbitrary in my opinion and to set one that could apply to all product types and data scenarios is likely not even understood."

What?

We wouldn't apply known thresholds with respect to some products because we couldn't apply that science across all products? Considering his rationale, the moment marketing orders were requested for a new product type, because we hadn't yet considered *that* product type, we'd cease applying applicable data to all other product types. For instance, if we had reliable data about lead exposure from combustible cigarettes, we could no longer consider that because we hadn't gotten a handle on lead exposure from ENDS. His rationale would systematically exclude our use of most science.

Considering our peach example, if someone applied for a marketing order for a new apple variety, we'd stop applying what we knew about peaches to our consideration of peaches.

He continued, "However, I think it's good you raise this question and are discussing it. Setting thresholds, likely would require rule-making, opportunity for public comment which are long processes, and tremendous resources of the center as well as legal analysis, and as such is not a priority given everything else."

As a toxicologist working for the FDA, I understood that this was a significant failure. But one didn't need that credential to understand that. And I never learned what priorities precluded something so fundamental to regulatory science, or why we couldn't use existing thresholds. It was also disturbing to first learn this after two and a half years with the FDA. Had there been a shift in our current thinking, or was this another unwritten policy? Instead of using science, why were we eyeballing data?

He wrote in respect to the entirety of our regulatory work, "No threshold is possible now or near future."

Threshold increasingly became a trigger word, with some leadership becoming so unhinged with its mere mention that it seemed they'd experienced aversion therapy. In mid-November of 2023, the three individuals who'd initially pushed fuzzy math violated policy by altering the record of a meeting where thresholds had been discussed. Leadership didn't respond when alerted. And in March of 2024, when misinterpreting a discussion about a data *range* as a threshold, Supervisor #4 neurotically snapped, "That's a threshold! We don't use thresholds!"

Did CTP not use *any* thresholds, at all, or was that just in the context of our work in the Environmental Science Branch?

On July 17th, 2024, in answer to the query, "Please explain how reviewers are to perform scientific reviews and assessments without using or considering thresholds," our Director of the Division of Regulatory Project Management wrote, "As CTP does not have the experience nor regulation due its creation in 2009, this is how we have operated. It takes time and experience to build thresholds that are predictable, consistent, and address the scientific need. Additionally, this is not a fast process—it requires CTP to follow Good Guidance Practices. Unlike other centers where regulation and guidance are in place, that is not the landscape in CTP/Office of Science."

Holy smokes!

That was a policy for the *entirety* of the Center for Tobacco Products!

Then what did we use?

Our context wasn't peaches. It was tobacco products. Our training said we should do credible work, as scientists, and thresholds were among the most informative tools we had to identify issues for further consideration. For example, if indoor emissions elevated exposures above a threshold where harm was likely, we needed to consider that and communicate that to the public. And we could.

And why didn't the "landscape" in the Center for Tobacco Products allow us to follow Good Guidance Practices?

And our external messaging consistently pointed to us doing something very different. Not only should reviewers have been alerted to this policy, but Congress and the public might've needed to know as well. Our leadership consistently misled Congress and the public.

13 | Put Words in Their Mouth

ANOTHER ISSUE HAD BEEN RAISED about false claims used in our assessments. The agency almost always based its regulatory determinations on claims made by the *applicant*, by the tobacco companies that applied for marketing orders. However, in some cases, applicants hadn't made those claims.

For instance, an assessment written by CTP might say, "The applicant stated that there would be no new or additional air emissions associated with use of the new product. Therefore, the agency determined there would be no significant impacts associated with marketing of the new product in the United States."

First, sorry tobacco industry, but your long and recent history of data manipulation and false claims should preclude any government agency tasked with protecting the public health from cutting and pasting any claims from applications submitted by the tobacco industry.

However, that's what we were directed to do. Many claims were cut, pasted, and massaged with no vetting.

But in some cases, the applicant hadn't stated what our assessments explicitly indicated they'd written! These weren't false or unvetted claims made by the tobacco industry, these were *our own* false claims.

14 | Weasel Wording

ANOTHER ISSUE HAD BEEN RAISED about the agency's use of information provided by the industry. In some applications for marketing orders, the applicant stated that the information they'd provided was *limited*. They cited COVID-19. Circumstances associated with the COVID-19 pandemic had limited, in some way, what they'd provided to the agency. This was information CTP needed to consider when making decisions about granting marketing orders for those tobacco products.

The statement was included in several instances in their application materials. Yet they hadn't indicated what information was limited or how. One instance of that statement framed the entire assessment provided, that I'd use to develop my assessment if we decided not to use theirs. It was in bold at the top of the first page of their assessment, so every bit of information that followed was subject to those unknown limitations. In essence, it said, "Here's our application to sell a new tobacco product in the U.S. Information we provide here for you to consider as part of your regulatory science determination for the protection of the public health mission is limited, but we're not saying how."

How would we handle this? Like the scenario where we'd been directed to use fuzzy math, we could've simply asked the applicant for a clarification.

CTP was looking for ways to expedite our work, and Congress had recently increased CTP's reporting requirements. So, the faster we could process applications, the better.

The Office of the FDA Commissioner had described our efforts in a 2015 analysis, "National Environmental Policy Act; Environmental Assessments for Tobacco Products; Categorical Exclusions." That analysis says, "roughly 120 hours [are needed] to conduct the assessments necessary to thoroughly evaluate an action's impact." The agency broke that 120 hours down. "CTP estimates that their tobacco experts invest roughly 112 hours (72 hours to prepare the assessment and another 40 hours to review it) while manufacturers invest 8 hours (8 hours to consult the FDA on certain subject matter)." It indicates some assessments might take less time, but some might take as many as 205 hours. It says that four hours are needed to prepare the FONSI, the document designating findings of no significant impact. Labor data they use from 2013 estimates a cost of about $120 per hour for that work.

Our work products might not have always been appropriate in a regulatory science context, but by 2021, the process was efficient. Preparation of the FONSI generally took about 10 minutes. It often took another 10 minutes for our Division Associate Director or Division Director to review the assessment and sign those findings.

The outcome of an assessment could be a FONSI, or an environmental impact statement (EIS) that is typically a much more time-intensive process. The 2015 analysis indicates that the "FDA expects to prepare one FONSI per assessment." So, the FDA didn't prepare for the possibility of an EIS. The agency had already predetermined FONSIs in every case.

So, where do you cut the fat in an already expedited process? Instead of preparing assessments ourselves, one means of expediting things would be to use the applicants' assessments. Having the tobacco industry prepare assessments that FDA scientists would otherwise prepare would save a lot of time and money.

In this case, in that context, I'd need to understand what we hoped to see in an applicant's assessment that we'd use, instead of writing the assessment ourselves. I asked my Supervisor #2 and we met on April 4th, 2023.

"I just thought," Supervisor #2 said, "we'd bring [Branch Chief #3 and #5] in on this conversation, just so we can see, you know, what we're all thinking about this paragraph that was inserted into the assessment. So, from my perspective, there's two things: 1) from when [Branch Chief #1] was here, she was pounding into our heads nothing editorial could be in the assessments and for me this is something editorial."

"Yes," our Branch Chief #3 and #5 quickly agreed.

They started that call focused on how we could disregard what the COVID statement said. They first said what it said should be disregarded because it was an *editorial* comment.

I looked up definitions more at FDA than in any other job. I wasn't usually looking up words with multiple meanings or complex definitions, and I'm not referring to scientific terminology. We didn't use flowery or highly technical language, by design, especially in our assessments meant to inform the public. Our training aimed to ensure our communications and work products were accurate and transparent, comprehensive yet succinct, and used clear language.

Instead, I needed to look up common words where their use in the workplace didn't align with my 50+ years in the U.S. as an English-as-a-first-language speaker. For example, when a document stated that "The applicant stated…" but there was no such statement, I'd look up what it meant *to state* something. I could see everything applicants provided and what we stated was stated was sometimes not stated. When leadership said documents that included inaccurate statements were accurate, I'd look up *accurate*. When our CTP Director #2 said a deceptive approach was valid, I looked up *valid*. When he said I acknowledged something that I didn't, I looked up *acknowledge*.

And *fraud* meant "an act of deceiving or misrepresenting." More than once, I looked up the meaning of *lie* that always still aligned with what I'd known all my life. Merriam-Webster says *to lie* is "to make an untrue statement with intent to deceive" or "to create a false or misleading impression." A *lie* is "an assertion of something known or believed by the speaker or writer to be untrue with intent to deceive," "an untrue or inaccurate statement that may or may not be believed true by the speaker or writer," or "something that misleads or deceives."

I'd been corrected when I chose a somewhat softer but equally accurate term when I reported what I was required to report about directives to use *deceptive* arguments. Our Office of Science Director #2 immediately recognized data had been intentionally misrepresented. I'd looked up definitions and thought carefully about the words that I chose, and in that case characterizing the arguments as *deceptive* was fair and accurate. However, he said that was unfair.

Unfair can be defined as "marked by injustice, partiality, or deception" or "not equitable in business dealings." Those didn't apply. Synonyms include words like *dirty*, *illegal*, *unsportsmanlike*, *foul*, and *nasty*. He couldn't have meant any of those. I found another definition, "not conforming to approved standards, as of justice, honesty, or ethics," that might apply if we remove "justice, honesty, or ethics" and inserted the truth about our workplace standards. What "approved standards" was I not conforming to when I reported what I was required to report?

To be fair, choosing the softer term was more than fair. For goodness' sake, we were a regulatory agency tasked with protecting the public health via our regulation of tobacco products. I looked up what it meant for Executive Branch employees to intentionally misrepresent information, lie, develop or use deceptive or misleading arguments, etc., and that failed to conform with standards of justice, honesty, and ethics.

Anyway, in this case I looked up the word *editorial.* When applicants had said things like, "There will be no negative impacts of marketing the new product because...", we'd previously excluded that. That was what Branch Chief #1 had wanted to exclude because it didn't seem appropriate to use the industry's opinions. What we were currently considering was different. The statement in question indicated information was limited in some way owing to access limitations. That statement suggested a *data quality* issue.

Supervisor #2 continued, "But what really bothers me about it is the way they phrased it as saying, 'This is all we know because there's nobody here, so this is all we can know.'" So, he did seem to understand that the information was limited in some way that we didn't understand.

However, he continued framing the issue as *editorial,* "And it, to me, it's kind of like *weasel wording* their way out of how to, if there's, out of I guess any ownership of if there's any problems there and what, and the reason why that bothers me is because when we publish this assessment, we're, because we don't have a guidance document, we can only say, 'Go look at assessments that are published on the FDA, on the CTP website.' And if applicants see these types of editorial comments, I'm just worried that, especially because they're so adversarial, they're not, it's not like Big Pharma, that they're gonna take advantage of this in some way and then start crafting these editorial things that pretty much relieve them of any burden and just put it all on us."

He was talking about how the agency didn't provide a guidance document with what we needed to perform our reviews and assessments. He referred to a more congenial relationship between FDA and Big Pharma, implying Big Tobacco was less cooperative. He was concerned that granting a marketing order with the statement visible in publicly posted documents would create a precedent where other applicants could use similar arguments.

However, the solution they'd identified was to not only prevent the statement from being *seen,* but to ignore what it *said.* The applicant had told us that the information they'd provided was limited, but not how.

He continued, "The agency is responsible for what's in the assessment and that everything that we want to be in the assessment is in the assessment. So, everything should be okay because the scientific part of it is there, and I get that, and I agree a 100% that we have to be on board with everything that's in the assessment scientifically and back that, but it's just this. It's just this *editorial* aspect of it that I'm worried that once it gets out there, that, and it's published, we publish it, that somehow the industry is going to take hold of these types of things and run with it. That's my concern."

Our Branch Chief #3 and #5 reiterated, "We are responsible for the contents of the assessment."

And Supervisor #2 repeated, "The scientific part of it is there."

What the heck was he talking about? *Scientific* means the principles and methods of science have been used, that evidence has been applied via a systematic approach. Therefore, it is not true that "everything should be okay because the scientific part of it is there." The applicant stated that the information was limited, but we didn't know if that was some or all the information they provided, or how that information was limited.

Responsible is a tricky one. Because the FDA is the agency taking action when it grants a marketing order, it is indeed the FDA's responsibility to adequately vet that action and communicate information about its decision-making process to the American people. Did they mean we have an *ethical* obligation, a responsibility to do things on the up and up? Or did they mean we *choose* what's in the assessment, that we are responsible for *selecting* what the public sees? This conversation points to the latter and contradicts the former.

What they said didn't align with my training, so I asked, "It's been a little while since I've been involved with considerations of applicant assessments. I think it's been quite a while. So, I was just wondering, what is the current thinking about what should be included and what should not be included?"

Supervisor #2 broke in, "I was just gonna say if this paragraph was in the SE Report, and not actually in the assessment, that would be fine."

What? That wouldn't be fine.

By "SE Report," he could've been referencing any or all the materials provided by the applicant, for example, including appendices, in its application for marketing orders under the SE application pathway. In this case, that didn't matter. The applicant's assessment, where the statement was included in bold at the top of the first page, was part of their application materials, and the statement was included elsewhere in their application materials. Because we were to make informed decisions, we were to review all information provided by an applicant, not just their assessment, because relevant information might be found outside of their assessment. Supervisor #2 was saying we'd have no concerns if the applicant had communicated the limitation elsewhere in their application materials, outside of the assessment.

Via his rationale, an applicant could include in their assessment a statement like, "There are no air emissions associated with the new product," but outside their assessment something like, "There are significant new and horribly deadly air emissions associated with the new product," and we'd ignore the latter.

Nonetheless, our Branch Chief #3 and #5 perkily agreed, "That would be fine, exactly. Mm hmm."

Supervisor #2 continued, "No problem with that at all."

Branch Chief #3 and #5 then said without pause, "But remember also even for typos in CFSAN they would correct the typos and they will ask the applicant to correct the typos in their assessment to that extent they look at the assessment and critique the administrative part. We're not able to ask these things on administrative part, so if we think in the administrative aspects of it there are certain things we don't agree or we have concerns and since we're not allowed to ask those from the applicant because it may delay the issuance of the marketing order. We cannot redact it under FOIA because there is no section under FOIA to redact that sentence that will be applicable to redact it. So anything that we do not redact from our document, there should be a section under the act FOIA that will actually fulfill that this information should be redacted otherwise it will be, we will be subjected to litigation if we redact something that's not really uh redactable by FOIA. So we cannot remove it and we're having the issue and I agree with [Supervisor #2] and we're trying, we don't have the guidance, and we're trying to provide really perfect assessments for the applicants in order to use and we still haven't gotten there and we still have certain things and aspects in their assessments that even sometimes administrative and we're not able to ask for it so it is safer and better at this time to prepare an assessment. The perfect assessment you asked about that is the one that really assess everything that we're looking at and at the same time doesn't have typos doesn't have legal arguments in it doesn't have all of these other administrative issues in it and that's when we use the applicant's assessment."

What?

I hadn't asked what was needed for the perfect assessment, and a lot of what she said raised concerns. She indicated another FDA center, the Center for Food Science and Nutrition (CFSAN), could produce more credible regulatory work products. This implied CTP's work products were less credible and therefore might not be credible. Instead of taking the responsible route that might entail reaching out to the applicant or simply not pretending that statement didn't exist, she indicated the solution could've been to remove the statement, that characterized the information as questionable, from public view. But we'd use the information anyway. However, she said we couldn't hide that statement if we used their assessment because FOIA provided no applicable category to do that.

In what respect and for whom was this *safer*? Not for the American people or for me as an Executive Branch employee who took an oath to do things on the up and up. And thank goodness FOIA didn't provide a category to redact statements that would reveal questionable decision making by a federal agency.

Further, she said we weren't allowed to ask for information from an applicant because that might delay a marketing order. We were unable to do credible work because that might delay the sale of a tobacco product.

She continued, "The Office of Science policy is not to, if there is a marketing order to be issued, not to delay that."

Holy smokes!

That was a policy I'd not heard before and that I'd never see in writing. The machinations required to make that happen violated agency training and our oath. But this was our current thinking. We were there to help move tobacco products on the market. Credible work couldn't delay that, and it seemed CTP was in the wrong agency. We served commerce, not the public health.

These two individuals were talking out of both sides of their mouths, indicating our work should be *scientific* while steaming ahead with an uninformed assessment that facilitated marketing orders allowing tobacco products to be sold in the U.S. However, I was still required to do credible work and abide by my oath, despite questionable unwritten directives.

In fact, trainings directed us to speak up in situations just like this. Those trainings indicated that was a critical role all Executive Branch employees had to support their agency's mission. One reason employees in roles like this needed permanent employment status was to ensure issues could be raised without fear of reprisal. Not only could the outcomes of potentially flawed agency decision making affect the public health in this one case, but questionable practices like this could compromise the credibility of an agency the American people needed to put their trust in.

So, I tried to refocus the discussion.

"It's quicker [for us] to do the assessment in this case," I said but, "I think, I was just thinking a little bit more broadly about things like, you know, when applicants provide additional information," because the applicant had said there was a data limitation.

Supervisor #2 doubled down and detracted, "But this one *editorial* thing just, is just, it's making me kind of cringe because of the way it's like, you know, a little bit of the legalese in there and I just don't want anybody to start grabbing a hold of that and yet, you know, if this was, let's say, SE and Chemistry had a question and there was going to be an AI letter, I would really push for, have them, since it's going to go to cycle two, have them just remove that part and resubmit it, we'd be great."

Branch Chief #3 and #5 confirmed, "Correct."

An AI letter was a letter the agency used to request additional information from an applicant. This was cycle one, the first round of reviews for these products. He said he'd push for raising our concern if another discipline like Chemistry had an issue. We could piggyback our requests with theirs. That would've allowed for a second cycle when we'd be able to evaluate responses the applicant provided to those requests. So, again, we'd need to pretend we didn't have a concern because no other discipline was raising an issue, not a valid reason a public health agency should use to avoid raising an issue that it otherwise would. And he was still campaigning for excluding the statement without considering what is said, how it applied to the information we'd use in our assessment.

He continued, "You know how Science Policy Branch gets? They're like, 'No,' you know, they don't want us wasting any time. This all came down from [Office of Science Director #1]. He was the one that was like, 'If you can do it in cycle one, do it in cycle one, and if you have to write the assessments, and that's the way it's gonna get done, then do it.'"

"Yeah," Branch Chief #3 and #5 agreed.

He continued, "Even though we were arguing, you know, 'That's not the way the other [FDA] centers do it.' So, I guess we're just going to have to pick our battles if, like I said, if this happened in an SE where we could go to a cycle two and there were other questions [from other disciplines], you know, maybe slip it in with that, but it's going to be sticking out like a sore thumb [where] we're the only discipline. I could see them saying, 'Can you write the assessment?' 'Yeah.' 'Okay, then just write the assessment.'"

She repeated, "Just write the assessment."

"And do you know, it's a *legal argument*," she continued, "it doesn't matter, it's not just surface, it's a legal argument that is in the assessment. Any legal argument in the assessment we really don't want it unless it relates to NEPA."

What?

Now a *legal argument*? The statement created a *data quality* issue, regardless of its potential characterization as a legal argument. Excluding it and ignoring what it said under that guise wasn't a credible approach. I focused again on the issue, "I'm looking at the statement, and even if we take it out, we still know it exists. So, we're still writing the assessment accepting this?" I repeated, "So, we see it, and we're taking it out for the public, but we're going to write the assessment knowing that?"

She ignored what I said and changed the subject, addressing Supervisor #2, "And I don't think Science Policy Branch will be in these meetings, yes, [Supervisor #2]?"

He indicated that they would be, "Uh no, they've been in every meeting that I've been in."

"They haven't been yeah," she responded as if she'd heard the opposite and continued, "I agree with [Supervisor #2]. I really have an issue with having it in the assessment and our assessment, it almost overrides their assessment because our assessment is focusing on the environmental aspects and we're not putting legal arguments there."

Our use in our assessment of their data, that was limited in an unknown way, didn't make that data no longer limited in an unknown way. So, I clarified, "This is the first statement in the, in their assessment, right?"

She agreed, "Yeah."

I read the statement that the applicant provided that indicated unknown data limitations, and I said, "So, again, I think that yeah, I just don't—"

She interrupted and was becoming increasingly pushy. She again detracted, "Our assessment adds additional information."

"I understand that," I said, but "we only know about the product what they provide to us and what they're providing to us, they've provided a caveat saying it's limited. So, whether we use—I think the question here isn't whether we use their assessment or not. If the concern is this COVID statement, it's the statement."

I further clarified, "The information we have comes from the applicant and the applicant is saying it's limited, and we're going to write an assessment based on something that we know is limited, and have it go through findings [when the agency makes a regulatory decision based on that information]. So, again, my point is that in this case, if we're looking at this statement, avoiding the statement doesn't solve the problem."

She rebuked, "But again our assessment is not based totally on their assessment."

I clarified, "No, but it's based on product information that they're saying is limited."

A few moments of silence passed before Supervisor #2 sheepishly chimed in, "Well, I mean, it could be limited, but they're still giving us what we need."

What?

"Yeah, they give us what we need!" Branch Chief #3 and #5 eagerly affirmed.

Their point was that the applicant had provided numbers that corresponded with the types of numbers we included in our assessments. This might be akin to getting a prescription from your doctor and the dose provided is a dose that might not apply to you, but it is a dose, so why not use it? The dose of that drug might be twice what you need or half, but it is a dose, so you take it.

Supervisor #2 and our Branch Chief #3 and #5 wanted me to cut and paste those numbers, regardless of that uncertainty, into an assessment with my name on it indicating the agency determined there are no significant impacts, so there was no delay in selling those tobacco products. "What we need" did not constitute credible information but was information that *appeared* to inform regulatory decision making. From an administrative perspective, it appeared to satisfy what the assessment needed to include. It allowed us to check the boxes. What they described doing was performative.

I asked "How?" that information satisfied what was required, was what we needed for a credible assessment and clarified. I said, "But, so, they're giving us what we need? But they have a blanket statement at the beginning saying that 'This is limited.' They're not telling us how it's limited. So, again, we're writing an assessment based on uh unknown, limited, limitations."

"Well, isn't that *every* assessment?!" Supervisor #2 enthusiastically posed as if he'd struck oil.

"Yeah!" our Branch Chief #3 and #5 confidently confirmed.

Supervisor #2 then added, "I mean, even *our* assessments are limited."

She again confirmed, "Yeah."

My training prevented me from celebrating their "Eureka!" moment and again required I clarify, "If we're writing the assessment because the COVID statement is a problem, because it shows that they're saying things are limited and we don't want other companies to use that and start picking up on it as a way to weasel out of providing more information, then it is indeed a problem for *this* assessment because we are writing an assessment based on limited information, that we don't know what's limited."

Supervisor #2 began, "You know—"

But Branch Chief #3 and #5 interrupted and was stern, "Again, assessment has to be administratively complete that we're going to be posting for the decision maker and our assessment informs the decision. It doesn't matter what the applicant assessment has in it in a statement. Our assessment informs the decision."

I started giving up and said, "Okay. I'll just, I'll write the assessment." I said, "I just don't, I don't feel like the point that I'm trying to make here is really getting through. And that's fine. I'll just write the assessment. I understand, not as good as you guys, but I understand reasons for why we do and do not do some things and all that. I'm just looking at this from the agency's perspec—"

Our Branch Chief #3 and #5 interrupted in an accusatory tone, "I want to ask you a question, actually. Some of those assessments that you talked about [previously] that have calculation and we don't agree with the calculation and we haven't been asking about that and we write our assessment. Isn't it the same?"

This seemed to be coming out of the blue and I wasn't following, "Can you, I'm sorry, can you repeat that?"

She explained, "I'm going through the same argument that you're doing. You're saying we're having the statement and we're preparing the assessment, we are brushing off the statement when we prepare the assessment, but when the applicant has said a calculation [in other applications] and it's not accurate as [Supervisor #2] said, in those, in some of those assessments for the applications and actually they not just even, they did calculation and they used model for drugs, they actually used a way for their assumptions and their calculation that is not even been used by other centers so we ended up writing the assessment. But that information is still in the applicant's assessment which we didn't ask about that."

"But this is a blanket statement for the entire assessment," I replied.

She was trying to convince me that a rationale previously used to exclude information provided by the applicant from assessments prepared by the agency in the past was the same. It wasn't. There were several differences. In those cases, CTP considered what the excluded information meant in the context of the assessments that CTP prepared. We weren't doing that now. And the current statement applied to all the information from the applicant we were to consider, but we didn't know how. That wasn't the case for those previous assessments. "Brushing off" the statement wouldn't get us off the hook.

I clarified, "That's the reason why [Supervisor #2] brought this up, they open the assessment with this statement saying, 'Everything that we're providing here is limited, but we're not going to say which parts of it.' It's the blanket statement that's provided to the entire assessment. If there's a part [like in the examples she mentions], it's different. It's similar, but it's not the same. No."

After an awkward silence, she said, "I don't know. I'm still more convinced that the assessment, preparing an assessment in this condition is much better than relying on the applicant's assessment with that kind of legal argument in it."

She was reframing the argument again, and we should've prepared the assessment. So, I said, "I agree with you, but my point is it's applying the blanket statement at the beginning of the assessment."

I again clarified, "It almost seems more responsible to say the applicant stated *the following* but we're going to go ahead and do *this*, but the applicant stated *the following*. I mean, I'm trying to be responsible from the agency's perspective here. I'm not trying to be a pain. I don't want to make more work for myself. I just am looking at what's happening here and saying, 'Hey, this is an interesting scenario and I'm trying to deal with it responsibly.'"

"Well, again," Supervisor #2 said, again contradicting himself and showing he understood the issue, "I think even with the blanket statement with them saying there may be limitations because not everybody was in during COVID, if the information that we *need* is in the assessment, that's all that matters."

Did he think I'd forgotten what he'd said just moments ago? We'd already debunked that attempted manipulation.

The statement suggested a data quality issue, regardless of our ability to prepare an *administratively complete* assessment that checked all the boxes. Excluding the statement and ignoring what it said—brushing it off—under that guise wasn't a credible approach. During this call he'd already described the statement as problematic. He'd critically paraphrased their rationale as, "This is all we know because there's nobody here, so this is all we can know.'" He'd even indicated it should be raised but that, "I guess we're just going to have to pick our battles if, like I said, if this happened in an SE where we could go to a cycle two and there were other questions, you know, maybe slip it in with that."

"Okay," I said, "We're going circular—"

He interrupted, "So, yeah, I mean, I know we're going circular but that's how I'm seeing it from our perspective. I guess from a scientific perspective, is that we're saying there was, you know, there may not, there may be information missing, there may not be, we don't know. Okay, well, we looked at it and we said, 'Well, everything that we *need* is there, so that's okay.' But I'm looking at it from the bigger, you know, two thousand feet view where I'm looking at the nefarious tobacco company trying to put some kind of, you know, limitation on it to limit their liability somehow."

She agreed, "Yeah."

So, why not just ask the applicant to clarify the limitation?

He continued, "But again, if we had any questions we would ask them."

However, he'd already said we couldn't ask. He'd already indicated that we had a question but that we needed to "pick our battles," that we couldn't ask because we were apparently the only discipline with a question. However, I'd never seen that limitation, where we couldn't ask, in any written policy and it violated written agency policies and messaging, and my training.

He continued, "In other words, let's say they didn't give us marketing estimates, we would ask that, and they had that blanket statement in there, okay, well, they missed that. That was one of the limitations maybe they were talking about, I don't know. We had to go back and ask maybe they know now, but they didn't know it at the time. So, we ask. So, I get both sides of it, but I'm looking at it from the big, again, the more of an administrative issue with companies that are adversarial. And I just worry about how if somebody gets the idea in their head, 'Oh, we can put all kinds of statements at the beginning of our assessments that somehow, again, you know, reduce liability.' That's what, that's how I'm looking at. But I understand completely what you're saying. Absolutely."

He again shows he understands but they keep rehashing inapplicable solutions. Having "everything we need" from an administrative perspective where we check all the boxes doesn't address the data quality issue. My training indicated we use sound science.

Bingo! She then mentioned timing. The agency had already decided to grant marketing orders for those products and leadership "wanted to move it fast."

Moving it fast wasn't something my training indicated should compromise the integrity of the assessment that I was hired to prepare.

From the perspective of a scientist required to do credible work, these sorts of conversations were incredibly concerning. And this sort of interaction wasn't rare. In fact, the amount of time spent infighting with a clear goal to quickly move a subpar product forward constituted a significant proportion of our work effort.

I reminded them, "I am putting my name on these documents and if I feel that I do not have accurate information, then I do not want to put my name on documents because, well, I don't have to explain why. We are the preparers and we've been hired to make judgment calls, and this is important."

I tried to further contextualize the problem, "We do our assessment and say, and this is kind of crazy, but say their marketing estimate, you know what they had at the time, they said 800,000 [units]. But it's really 800 billion. Would that potentially be a problem? So, say it's like 'Okay, that's ridiculous.' Well, I don't know what I don't know. If we're making an assessment based on marketing estimates and we don't know which parts of this assessment are potentially, you know, wrong, then I'm not comfortable doing it."

Our Branch Chief #3 and #5 again mischaracterized the issue, "I recommend bring in this legal argument, that you saw this legal argument, at the preliminary assessment meeting and let CTP deal with that."

Reviewers shared potential concerns with the Regulatory Health Project Manager that coordinated logistics, the Technical Project Lead that considered scientific considerations across all disciplines, our Science Policy Branch, and reviewers from other disciplines at preliminary assessment meetings.

I said, "My role is to evaluate this information—"

She interrupted, "Of course."

I continued, "—and with this statement there I can't consider this information to be accurate. I will go ahead and start preparing the assessment because that's what I would do anyway, just so, it's like, everything's documented. But this isn't the sort of thing that, I don't think, I think it would be irresponsible to not bring it up."

She again said, "Put it in your preliminary meeting notes."

I agreed, "I'll do that."

She continued, "—that, 'We don't have questions, but here is, we found the statement.' Put it there in the preliminary meeting notes."

But we did have a question. When we said "we don't have questions" that meant we had nothing to raise, no issues we needed resolved.

So, I explained again, "The information that I need, in order to do my evaluation, there's a caveat, and based on this caveat, I do not know the accuracy of the information that I'm looking at. So, I need the applicant to look it over and tell me what parts are accurate, and what parts are not."

I again reiterated, "The accuracy of the information I'm to evaluate, to do my assessment, is questionable because of this statement."

Supervisor #2 asked, "Okay, so would a statement coming back from the applicant stating that, you know, 'We reevaluated the assessment and everything stands, it looks good to us,' you know, 'We talked,' you know, whatever, you know, with a statement, getting a statement from them, you know, alleviate the issue?"

I said, "I think so. Yeah. I mean if they came back and said, 'Everything's accurate,' or they came back and said, 'Now we're going to update this number,' yeah. I just, because I mean if I'm going to do an assessment, I need to have confidence in the information the applicant is providing. And if they come back and say, 'Remove the statement. The statement's no longer applicable. This part of it is, was, accurate. This part was an estimate,' you know and I could even say, 'Please provide information about which aspects of the information that you've provided is, uh, is accurate. What's estimates or no longer apply,' something like that. It could be very succinct."

The plan leaving that meeting was to include the issue in the preliminary meeting notes and see what happened.

15 | Good Cop, Bad Cop

SUPERVISOR #2 CONTEXTUALIZED LEADERSHIP BEHAVIOR and our subpar work immediately to follow in a second call on April 4th, 2023. Our Branch Chief #3 and #5 was not present, and his tone was different.

He said, "I think we have a really good plan going forward and I'm kind of excited."

He said, "I understood everything you were saying, even when you got interrupted. When you were able to come back in, all your thoughts were still coherent. I understood everything and I think [Branch Chief #3 and #5] did too."

He said, "I completely understand and respect your perspective and I think that's fine. We need to bring it up and say, 'They're telling me not everything might be here. So, let's go back and ask them, "Is everything here or not? It's two years later and you've got all your people there."' I think that's valid. I really do think that's valid."

He said, "There might be ways that we can start strategically working on small things, work on the small victories, like, maybe in this way we can ask the question and Science Policy Branch starts hearing us."

He said, "I'm going to completely back you on this. If [our Deputy Director for Regulatory Management] says, 'Well, [Supervisor #2], we can't do this,' it's like, '[Deputy Director for Regulatory Management], it says this might be limited. We just need to know if, since this was written, yes, this is all there was. And if they say, "Oh yeah, by the way, that's all there was now that we looked it over again, and that's it," good, good to go, that's all we need. But they put the statement in there, [Deputy Director for Regulatory Management].'"

When he said, "We can't do this," he referred to our inability to ask the applicant a question, that our Deputy Director for Regulatory Management might say we can't resolve a data quality concern.

He also said, "I'm sorry that was your first full-blown [Branch Chief #3 and #5] stepping on your conversation. But I've seen it. I've been here eight years and it's been what you saw, unfortunately. I hate to say it. It's pretty normal."

Our training indicated that our ability to freely raise concerns and do credible work were paramount, and her behavior countered that. So, I asked, "Do you think that affects what we, does she do it, like, and I don't, I'm not trying to blame or, I'm just trying to understand. Do you think she does it in ways, like with Science Policy [Branch] and such that might affect our ability to get things through?" I asked because she tried to script me to tell them, "We don't have questions," when we did.

He said, "I really don't know." And he said it again, "Again, I'm sorry about [Branch Chief #3 and #5]. Maybe you haven't been in very many one-on-one meetings with her. That's the way it goes. I get stepped on all the time."

He'd swung back and forth between credible and compromised work when Branch Chief #3 and #5, who was his direct supervisor, had been present. Now he said, "[Branch Chief #3 and #5] was just like, 'Let them deal with it.' I'm like, 'No, that's not—.' Let them deal with it? But they still need to hear us, because I don't want them to sweep it under the rug and go, 'Just do it.' No, that's, there's another issue here, and that is that they said they didn't put everything, they may not have put everything in there that needs to be in there. Well, now we need to go back to them. And I don't want them to sweep it under the rug and say, 'We'll take this under advisement,' or 'We'll file it away,' or 'We'll talk about it,' and you know what happens. They'll never get back to us on it."

When he said he didn't want leadership to sweep it under the rug and say, "Just do it," he referred to the expectation that we ignore what the statement said and write an uninformed assessment that supported marketing orders. He continued, "I'm hoping it doesn't just wither away and we get that, 'Well, we'll have to talk to ELT and we'll get back to you.' Because that's not what I want to hear. I want to hear, 'Yeah, that's a pretty simple approach. Hey, RHPM, can you set up a call? Just give 'em a call and say, "You put the statement in there. Is there any new information or only—that's all you got?"' Hopefully, [our Deputy Director for Regulatory Management] will be amenable to that, the TPL will be amenable to that, and they can make it happen."

ELT was the Office of Science's Executive Leadership Team that included our Office of Science Director #2, our Deputy Director for Regulatory Science, and our Deputy Director for Regulatory Management. The RHPM was the Regulatory Health Project Manager that coordinated logistics for CTP's review of those products. And the TPL was the Technical Project Lead coordinating

CTP's scientific evaluation of those products across all scientific disciplines. The statement applied to information considered by all scientific disciplines.

He continued, "I don't want to hear from Science Policy Branch, because Science Policy Branch is always the one [that says], 'Well, if you have everything you need, then you can do it.' We don't know if we have everything we need, because [the applicant] said, 'We don't know if we put everything in here. Everybody wasn't at work because of COVID.' So, we could throw that back at them, but they still push us and say, 'Yeah, but everything's in there. You need to write your [assessment],' and I'm going to have to leave that—I'll back you and say, 'Look.'"

He indicated ensuing meetings would be contentious. He said, "So, just be prepared to be cross-examined harshly from Science Policy Branch, possibly [our Deputy Director for Regulatory Management]. I'll back you and say, 'Look, none of these places have ever put these statements in. So, we have to take it at face value that everything they knew was there. But they put this statement in saying, "Not everybody was at work. We did the best we could with who was available to give us information, and this is what we came up with." Okay, well, times are different now. We're looking at this now.'"

He continued, "And you need to confirm, and I'll back you 100% on that. And [you] just have to be prepared for the cross-examining: 'Can you do this?' Because that's what they're going to do. They always do it. 'Can you do this? Can you do that? Why can't you do this? Why can't you do that?' That's the type of thing that they do."

"We're going to have to be prepared for those types of questions," he repeated. "It's just like, 'If you have everything you need, then you don't need to ask them any questions. It doesn't matter what their statements are. You have everything you need.' Well, we don't know that we have everything you need. And, you know, so there might be new information. We don't know."

The circumstances were easy to explain and the logic was simple. That was usually the case, but the drama in situations like this was often over the top. Leadership behavior often ranged from anxious to neurotic when faced with the prospect of communicating what our training required we communicate. What did they fear? Were they getting their nuts zapped through their government issued equipment? Had they had toes severed and eyelashes ripped out? Had they found their dogs dead on their doorsteps?

I said, "You know, if I'm going to do an assessment and that statement stands, then my assessment is going to be much more of an estimate than a reflection of data." I said, "It could have been somebody's toddler making up the numbers or something even, right?"

I said, "As a responsible, you know, federal employee that wants to do a, you know, a good job with this assessment," but we didn't do a good job, so I continued, "*Good* is not a good word for that. *Good* isn't, it's like, just, you know, you know, to make sure that we're *compliant.* I need to make sure that the information I'm evaluating is *applicable.*"

He referenced groupthink that didn't align with our training. He said, "We need to show that there was an outcome—good, better, or otherwise—but that we went in as a team and we went for it and you're going to win some, you're going to lose some kind of thing. But we need to show, bring it up, everybody gets on the same page and off you go and you show a unified front."

He said, "Something that [our Branch Chief #1] talked about, at least when I first started, I don't know if she was saying it to everybody, but that the most important thing is that we always show a unified front, we're not saying something different. And the reason is because, believe it or not, there are people within CTP that would come to us and try to get a different story from different people. They'd conquer and divide and conquer, kind of thing, and so we always made sure that we were showing a unified front."

What?

My training indicated that we could disagree, that disagreement was good, and that allowing disagreement among scientists was paramount. And if we handled those disagreements with integrity, why would we need to pretend we all agreed? What was the big deal?

And who was dividing and conquering? And why?

Our training required that we raise issues, and he said leadership frequently hadn't done that. He said, "[Our Branch Chief #1] didn't like to always rock the boat and she tried to pick her battles, but sometimes picking the battle was like no battle at all. And even though some of us were like, '[Branch Chief #1], we should really bring this up in a preliminary assessment meeting and talk to policy and see if we can slip this one in.'"

Why were we always needing to slip in the science?

He explained what he'd said during the previous call in the context of litigation, "I'm looking at it from the, again, more of an *administrative*, *procedural*, trying not to give the adversarial companies any kind of advantage, because they are already trying to hamstring us at every turn and corner. You know that they can. And it's really maddening, especially coming out of Pharma, where everything was very cooperative. We didn't always like it, and sometimes we'd argue and say, 'We don't think we need that study because, blah, blah, blah, blah.' And sometimes they would agree, and sometimes they wouldn't. But it was never really more than scientists talking. This type of adversarial—it was never this. It was never, 'We're going to sue you.' We would never, ever do that because then, well, it just wouldn't happen. So, at any rate, so that's the perspective, about, I was looking at it. And so, I understood your perspective completely."

But, in this case, what would they sue us for? We could ask applicants for more information or for clarifications. It didn't seem that would warrant a lawsuit.

Our work on those applications was then put on hold.

I later learned that CTP had previously ignored similar statements in other applications. Leadership knew that during those April 4th, 2023, meetings. They weaseled around even having a discussion about that, instead saying, "It's *editorial.* It's a *legal argument.* It's an *administrative issue,*" and that a performative assessment that *appears* administratively complete would suffice—more rubber-stamping.

Other applications included a statement indicating that information was limited, but CTP hadn't bothered asking, "How?" Asking now would mean a change in course that would call into question previous regulatory decisions and risk litigation, fairly, from the industry and maybe from public health groups. Unwritten policy increasingly indicated that avoiding litigation, and maybe other things, took precedent over credible work and the public health.

On November 14th, 2024, work on those applications resumed. Supervisor #2 had been replaced in January of 2024. By November of 2024, Supervisor #7 was at the helm. She pressured me to sign a review that lacked any details about what we'd considered and that indicated we had no questions or concerns.

The Regulatory Health Project Manager managing that assignment asked me to review information about those products, considering the time that had passed, to ensure our considerations were up to date—it had been a year and a half. So, I was required to do that. And I'd need to learn if and how those data quality issues had been handled. On that list of *written* duties, it said each primary reviewer "ensures accuracy of uploaded reviews and provides corrections or addendums to respective reviews, if discrepancies are identified." This was a normal sort of thing to do, and I needed to follow his directive.

However, when I tried to do that, Supervisor #7 wrote me up. She lied about what happened, and because all those communications were in writing, I provided that evidence. She and our leadership ignored that, and then, on January 22nd, 2025, she wrote the same already disproven lies in my annual performance evaluation. She wrote, "I would like to see improvement in efficiency. Dr. Leppanen has had issues with this, for example finalizing the [product name] review on Tuesday 11/19/2024. A task which should have taken no more than 30 minutes, took 3 hours with multiple confirmation emails to finalize the review." That was her only example, and it was a lie.

She had absolute immunity, and I still had to follow the rules. So, I was punished for following a directive to ensure our work was credible. I was supposed to, but unable to, ensure our considerations were up to date. Using coercion, leadership again lied and compromised the credibility of our work to cover up past indiscretions and facilitate marketing orders.

16 | Leaders are Above the Law

I FOUND A FLIER ON MY WINDSHIELD, snapped a photo, and sent it to a colleague when I recognized things that increasingly happened at work.

Characteristics of a Cult
Are You Brainwashed?

Authoritarian Control
"Group think" requires that all agree.
Independent thinking is not allowed.
Questioning is not allowed.
Behavior is aligned via strict rules.
Threats, shame, and guilt are used to manipulate.
Propaganda pushes the party line.
Isolation
Ties with outsiders are cut.
Outside information sources are forbidden.
The outside world brings about fear and paranoia.
Those who leave are shunned.
Elitist
How we do things is superior.
Only we can provide salvation.
We know the truth.
Only we know the whole truth.
It is us versus them.
Leaders
Leaders provide no transparency.
Leaders garner excessive loyalty.
Leaders provide the only valid source of truth.
Leaders are always right.
Leaders are not held accountable.
Leaders are above the law.

17 | Following the Road of Yellow Brick

THE CTP OMBUDSMAN SEEMED FLUSTERED on January 4th, 2024, after I'd reached out to raise issues further about events surrounding the fuzzy math incident. He repeated that the inaccurate Scientific Dispute Resolution Determination was "issued" and "final." We both knew issues could and should be raised further. Inaccuracies in that determination and other serious concerns he knew about kept falling off the table.

How the CTP Ombudsman and Associate Ombudsman handled things had been inconsistent. Their behavior initially seemed genuine. They'd both said they were concerned with evidence of wrongdoing and with the broader workplace culture that they'd described. They said staff feared going on the record, so they encouraged me to raise concerns. However, the CTP Ombudsman said discussions were confidential, then shared by name leadership too afraid to go on the record about concerns, some similar to mine that hadn't been addressed.

Had it caused problems when I raised issues already known but ignored by those higher in the hierarchy?

The Associate Ombudsman also talked about previous attempts by staff to raise issues. She said those attempts didn't go well and that opinions of staff that raised issues were not good. She said, "That was in [Office of Science Director #1] days"—before he'd left to work for the tobacco industry.

"I don't know if you knew [him] very well," she said, "but he took that very—not great—because it was [his] way or the highway." She hoped our new Office of Science Director #2 would be cooperative.

She'd also mentioned entering information into the administrative file. So, after our CTP Director #2 issued his inaccurate SDR Determination, I asked about submitting clarifying documents into that file. This would ensure that information excluded from his SDR Determination would inform any subsequent considerations. Inaccurate statements in his SDR Determination should've also been noted there.

According to the *U.S. Code of Federal Regulations*, Title 21 § 10.70, "FDA employees responsible for handling a matter are responsible for insuring the completeness of the administrative file relating to it." The National Archives describes the *U.S. Code of Federal Regulations* or *CFR* as "the codification of the general and permanent rules published in the *Federal Register* by the executive departments and agencies of the Federal Government." The *Federal Register* is the daily journal of the U.S. federal government and contains "federal agency regulations, proposed rules and notices of interest to the public, executive orders, proclamations, and other presidential documents." It "informs citizens of their rights and obligations" and is of interest to those "concerned with government actions that affect the environment, health care, financial services, exports, education, or other major public policy issues."

The FDA described there how we were accountable to the American people who should be able to trust what's described there actually happens. As a federal employee, I needed to understand and follow those rules. 21 *CFR* § 10.70 also said, "FDA employees working on a matter have access to the administrative file on that matter, as appropriate for the conduct of their work."

21 CFR § 10.70 Documentation of significant decisions in administrative file

(a) This section applies to every significant FDA decision on any matter under the laws administered by the Commissioner, whether it is raised formally, for example, by a petition or informally, for example, by correspondence.

(b) FDA employees responsible for handling a matter are responsible for insuring the completeness of the administrative file relating to it. The file must contain:

(1) Appropriate documentation of the basis for the decision, including relevant evaluations, reviews, memoranda, letters, opinions of consultants, minutes of meetings, and other pertinent written documents; and

(2) The recommendations and decisions of individual employees, including supervisory personnel, responsible for handling the matter.

(i) The recommendations and decisions are to reveal significant controversies or differences of opinion and their resolution.

(ii) An agency employee working on a matter and, consistent with the prompt completion of other assignments, an agency employee who has worked on a matter may record individual views on that matter in a written memorandum, which is to be placed in the file.

(c) A written document placed in an administrative file must:

(1) Relate to the factual, scientific, legal or related issues under consideration;

(2) Be dated and signed by the author;

(3) Be directed to the file, to appropriate supervisory personnel, and to other appropriate employees, and show all persons to whom copies were sent;

(4) Avoid defamatory language, intemperate remarks, undocumented charges, or irrelevant matters (e.g., personnel complaints);

(5) If it records the views, analyses, recommendations, or decisions of an agency employee in addition to the author, be given to the other employees; and

(6) Once completed (i.e., typed in final form, dated, and signed) not be altered or removed. Later additions to or revisions of the document must be made in a new document.

(d) Memoranda or other documents that are prepared by agency employees and are not in the administrative file have no status or effect.

(e) FDA employees working on a matter have access to the administrative file on that matter, as appropriate for the conduct of their work. FDA employees who have worked on a matter have access to the administrative file on that matter so long as attention to their assignments is not impeded. Reasonable restrictions may be placed upon access to assure proper cataloging and storage of documents, the availability of the file to others, and the completeness of the file for review.

The Associate Ombudsman had said that a memo in the administrative file could capture what had happened with the deceptive approach, that "anyone who FOIAs documents for their product would see it plain as day." She didn't say if others making Freedom of Information Act requests would see it.

"It wouldn't be an overwrite memo," she said, "and it wouldn't be anything that would rescind an order, but it would be captured in the administrative records." Marketing orders wouldn't be rescinded. Use of the deceptive approach would be noted in some way but not corrected or considered. However, documenting what had happened in the administrative file might prevent the same or similar sorts of things from happening again.

If reviews needed to be corrected, she said, "That would be more so like a formal process because of the involvement of every single center within CTP having to be roped into it." Rescission letters, especially because FDA scientists intentionally misrepresented data, risked litigation and possibly more if discovered by the public, who those assessments were meant to inform. CTP would need to indicate why those letters were issued. Further, redos would involve all scientific disciplines to ensure considerations aligned with current practices. That would further delay the process and risk litigation. Time needed to redo reviews of possibly thousands of tobacco products would delay reviews of other tobacco products, which also risked litigation.

The CTP Ombudsman mentioned filing those clarifying documents instead in my *employment* record. He oversaw implementation of the SDR process. So, he might've been disappointed its implementation might not be rectified under his watch in CTP. Performance requirements said issues should be handled at the lowest administrative level. The next step would elevate concerns outside of CTP to the Office of the FDA Commissioner.

However, entering clarifying documents into my *employment record* would in no way inform the *administrative file*. What a strange idea! I wanted to confirm that's what he meant. His behavior was weird and seemed a little threatening. He must've meant the *administrative* file, not my *employment* file.

"When I read about 21 *CFR* § 10.70 and employees' ability to access the file related to their job, that's what you're talking about," I asked, "that's where it would be accessible?"

He evaded, "Ah, if you have an *employee* file somewhere, *maybe*?"

What?

Again, this was weird. So, I refocused, "No, not an *employee* file, the *administrative* file related to the SDR issue."

He then changed gears and said, because the issue might not be considered *significant*, those clarifying documents might not be entered in the administrative file. "Right," he said, "and I believe 10.70 mentioned, you know, a *significant* impact or a *significant* decision from the center, so, then the interpretation, I don't know who defines *significant*. Your determination may be different from mine which may be different from [the Associate Ombudsman]'s, you know. I don't know how that goes but there's not necessarily like a jacketed file somewhere where we keep these things."

He'd previously been articulate and knowledgeable about his role as our CTP Ombudsman. Now, he seemed befuddled.

21 *CFR* § 10.70 was prefaced with, "This section applies to every significant FDA decision on any matter under the laws administered by the Commissioner, whether it is raised formally, for example, by a petition or informally, for example, by correspondence."

The problem with his current argument was that the CTP Director #2 had already determined the issue was significant enough to direct to the administrative file. His inaccurate SDR Determination was there, wherever that was, freewheeling and unjacketed. But it was someone's responsibility to make sure the administrative file relating to this issue was complete. If his inaccurate SDR Determination sat there alone without the SDR Initiation Memorandum that came beforehand and without these clarifying documents, those responsible for handling the matter were very knowingly not ensuring the completeness of the administrative file relating to it.

In reference to documents entered into the administrative file, 21 *CFR* § 10.70 also said, "Once completed (i.e., typed in final form, dated, and signed) [those documents could] not be altered or removed." It also said, "Memoranda or other documents that are prepared by agency employees and are not in the administrative file have no status or effect." By cherry-picking what moved into the administrative file, CTP would again misrepresent information using a deceptive approach.

Every level of leadership ignored and discouraged requests to access our procedural documents, the administrative file allowed by 21 *CFR* § 10.70, or any other administrative record or file associated with our work. And we also needed confirmation about our role, if we were required to follow directives to use unscientific, ignorant, and misleading arguments in our regulatory work that were increasing in frequency.

Where was our process described? We hoped to find that there.

Our Director of the Office of Science's Division of Regulatory Project Management did finally respond months later. On July 17th, 2024, she wrote, "For NEPA compliance you would have to review every single action where NEPA is required (i.e., all applications or petitions requesting agency action). As a reviewer and Office of Science employee, you have access to all applications through our archive. Therefore, you can utilize CTP Search to view the agency's decisions regarding NEPA on that action. To give you a current sense of applications—you would need to search the following (note that my numbers are provided as of 6/30/2024): 3,849 closed EX Requests, 9,813 closed SE Reports, 26,097,135 closed PMTAs, 49 closed MRPTAs."

During an October 24th, 2023, public meeting "Premarket Applications: Opportunities for Stakeholder Engagement," our Office of Science Director #2 said staff efforts to process those "26 million applications over the past few years was quite a Herculean effort," and a lot of us certainly wondered how so few staff had accomplished that.

But instead of pointing to procedural documents with sound approaches we'd follow, our Director of the Division of Regulatory Project Management directed staff to review millions of documents, a Herculean effort. And like our other leadership, her external messaging differed. For example, at that same October 24th, 2023, public meeting, she told the tobacco industry, "Our goal is to be transparent with our decision making, discuss some of the science and how it's changing and how that factors into our determinations."

So, maybe the industry could help CTP staff understand what factored into our determinations. They didn't want us to see the record, and they didn't want us to go on the record. And it was never confirmed, as per 21 *CFR* § 10.70, that the "FDA employees responsible for handling [this] matter … [insured] … the completeness of the administrative file relating to it."

When I reminded the CTP Ombudsman that the SDR Determination was inaccurate and excluded relevant information, he said, "[Our CTP Director #2] doesn't, you know, whether he heard all of that conversation or not, it's beside the point, what he made a determination on was what he *heard*."

What?

What he *heard*? Why had he stressed that word?!

"Beside the point?"

Was their goal to be accurate and use that SDR process as described or to build a case to support a certain outcome? It was the latter. CTP consistently cherry-picks to build a case for its desired outcome.

Among other things, our CTP Director #2 left out any mention of the already implemented solution. That solution that had already been implemented had been implemented because the deceptive approach had already been determined to be invalid. His inaccurate SDR Determination listed meetings with individuals who knew about "all of that," including our Division Director and Office of Science Director #2 who came up with the solution. And I shared "all of that" with him as well.

Here was another individual—this one tasked with implementing our SDR process meant to protect scientific integrity—pivoting sloppily to accommodate another coverup by essentially saying, "Oh, we don't keep records. I'm the CTP Ombuds in charge of implementing this process, and I don't know what *significant* means, but *this* isn't that. But we can document this in your *employment* file. And facts? Those are beside the point."

What had our CTP Director #2 *heard*, and from whom? With all this in mind, it struck me when I read his SDR Determination that he might not have written it. I recognized language used by complicit branch and division leadership. Our

CTP Director #2 was a busy person and had never done the sort of work involved. Others more familiar with the issues would no doubt have assisted.

To cover their butts, it looked like others might've misled our still fairly new CTP Director #2. And acknowledging that the already implemented solution even existed might counter the validity of the deceptive approach that the solution replaced, and those previous reviews and assessments might need to be redone, and that would risk litigation. So, a different narrative was needed.

Had he *heard* an inaccurate account of things? Had they taken advantage of his naiveté? Had he signed the inaccurate SDR Determination believing it was on the up and up? Or had he known this was a means to silence a slew of issues that industry, Congress, and the American people might find concerning?

He'd seemed sincere, but he knew the meeting we'd had wasn't accurately represented in his determination. He'd acknowledged receipt of consequential information excluded from his determination. He'd ignored written documentation he'd known about and had been offered. The CTP Ombudsman and Associate Ombudsman likewise knew his SDR Determination was inaccurate and that other serious issues were ignored.

Regardless, he signed that determination, so he owned it.

People were still lying and it was still my job to raise those concerns. And each additional interaction increasingly pointed to broad, chronic, and intentional wrongdoing. How could that cascade to affect the public health?

The next step in the process was the FDA Office of Scientific Integrity. When issues were communicated there, the FDA Ombudsman wrote that the issue involving the deceptive approach "was not ripe" because the deceptive approach was no longer used. So, he understood there'd been a problem for which a solution had been implemented. Other issues remained, like a flawed record, dishonest leadership behavior, and, alongside our credentials, we still used unscientific, ignorant, and misleading arguments.

The FDA Ombudsman said issues should be handled in CTP, and my job was to follow the process. So, back I went! I went back to our CTP Director #2, who directed me back to our Office of Science Director #2, who directed me back to our Division Associate Director. In the meantime, our Division Associate Director wrote that he agreed with the inaccurate SDR Determination, confusing staff further because the new approach had already been implemented.

18 | Conversations with Nathan Thurm

ON MARCH 20TH, 2024, our Office of Science Director #2 wrote, "I'd recommend setting up a meeting to talk with [the Division Associate Director]. Let me know how that goes."

Our Division Associate Director subsequently ignored requests for, canceled, and changed the topic of that meeting. His behavior was consistent with his previously documented behavior that our Office of Science Director #2 knew about. However, I was directed back to him, so I couldn't drop the ball. I needed to follow the process.

I'd let our Office of Science Director #2 and our CTP Director #2 know, "[He] has been looped in or directly contacted about all of these concerns and generally ignores anything but easy fixes (e.g., timecard issues), but I'll give it a try. Yesterday during our branch meeting he said in reference to delays in attention to our branch's work, that if he had a magic wand he might be able to get to our work sooner. This sounds like petty reporting, but until management takes NEPA seriously, staff will have a hard time supporting compliance."

Our Division Associate Director was the primary individual tasked with direct oversight of our review and assessment process. On November 15th, 2023, he'd written, "Our process for all branch staff is for me to receive a clean assessment (i.e., watermark, comments/track changes removed) and corresponding FONSI work products after the primary and secondary reviews are deemed complete by the reviewers that the documents are ready for me to review. In your role as primary reviewer, you would email the clean assessment and FONSI to me coping your secondary reviewer indicating the documents are ready for me review."

He'd overseen, reviewed, and made agency determinations informed by the deceptive approach and all the other approaches used. He met with our team supervisors and supervised our branch chiefs. And he stepped in to fill vacant roles as my Team 1 Supervisor #3 and our Branch Chief #2 and #6. He'd often served as the decision-making authority when input was needed, or he consulted with our Deputy Director for Regulatory Management or others on our behalf. Our Division Director was his supervisor and occasionally stepped in for him.

We needed to have that meeting to clarify a few things, and several issues slated for discussion were critical. Staff needed to know if those of us preparing assessments needed to agree with the arguments posed in those assessments. Leadership below him talked out of both sides of their mouths about that and wouldn't address the issue directly. He'd ignored that topic in written communications, and leadership above him sent us back to him. We needed to know if work products, especially where our credentials were listed, but regardless, should align with our credentials, with what we'd been hired to do, with agency messaging, and with our training. If leadership again told us to do something that violated our oath and training, and we did it without evidence that we'd been directed to do that, we'd be culpable for violating our oath and training.

This was a bad scene.

Were we expected to violate our oath and training by following unwritten guidance from leadership with directives to never find any issues with the tobacco products the agency was tasked with regulating? Was this why we didn't put how our assessments were developed and reviewed in writing and why the administrative file was unavailable? Regardless, verbal or written instructions to do anything that violated our oath and training required I raise those concerns via the processes made available to do that, regardless of how circular and unproductive those processes were.

Staff also needed to know if we'd revert to the deceptive approach that had been replaced or use any other deceptive approach that misrepresented data or other information. A solution had been implemented to prevent that one fuzzy math tactic used to misrepresent data. However, staff were subsequently confused by the inaccurate SDR Determination that now indicated use of deceptive arguments was valid, and that contradicted public messaging coming out of CTP about the high standards we followed.

Writing to our Division Associate Director about these topics hadn't been successful. When he did respond in emails, he often wrote strange answers that didn't apply. The Associate Ombudsman had recommended writing short requests to increase the odds those pressed for time might read what was communicated. So, that's what I did.

For example, we were excluding a lot of information that we knew about from our assessments. So, we'd asked, "Could we please learn about any rationale for excluding information from the assessment?"

He responded, "The point is to focus on our energies in one place first and see what information is found and discuss it. At the CDC/FDA Fireside chat with Commissioner Califf this past Tues he emphasized we should focus our energies. Taking on all the references in the entire assessment would be overwhelming and go too many different directions as a first assignment. This is a straightforward literature search and discuss the data that is found assignment. Air quality section was what I assigned only under that rationale. Nothing beyond that for now. I would like to see what was found. But it requires analysis of the literature and scientific discussion. As scientists, this should be an enjoying thing to do and should be a collaborative activity, where we respect others' opinions."

What?

He'd avoided the question and chose to discuss other things.

That question had been raised a few times. A variety of exposures were excluded from our assessments, and the reasons for that were unclear. If we understood those reasons and the topics that should be excluded, we might not expend effort trying to add that information. There must've been reasons. For instance, why did we exclude a considerable amount of relevant information about secondhand and thirdhand emissions?

During the same exchange, I'd written, "Management should keep in mind the expectation that we risk our professional reputations as preparers of the assessment where our names and experience attest to the credibility of these public documents. Using our credentials without our agreement is questionable."

In response he only wrote, "Acknowledged, we work as professionals." But we weren't working as professionals, and these responses didn't address what were important issues fundamental to the agency's regulatory compliance. And this happened a lot.

We'd been asked to update our assessments and had learned that industry-reported market trend data used in our assessments contradicted other possibly more credible data sources. We were using data the tobacco industry reported about taxes to the Alcohol and Tobacco Tax and Trade Bureau, or TTB data. But elsewhere in CTP, even elsewhere in our own branch, we used data from Nielsen or Euromonitor International, in part owing to concerns about self-reported data from the industry.

During a Team 1 meeting on February 20th, 2024, one colleague had just finished mentioning an effort to organize information that we were summarizing in a spreadsheet and then asked, "Are we set on using TTB data or is it okay if we want to consider some other sources?"

In the context of that discussion, we needed to know if we'd continue using TTB data or use those other sources that the rest of CTP regularly used. This was important because the agency's findings of no significant impact were supported by TTB data that showed a steady or declining market. Other sources indicated an increasing market for the same products and therefore didn't support the agency's findings of no significant impact.

As the decision maker for the agency, our Division Associate Director who was also our Team 1 Supervisor #3 at the time was absolutely the appropriate person to ask. He'd been briefed previously on several occasions, with sufficient detail, about that specific concern. It really looked like we were again cherry-picking data to support findings that would facilitate marketing orders. We wanted to be sure we used the most credible data available and developed credible arguments, as our oath and training required.

However, he responded with only, "Yeah, I think that's a great question." He chose not to answer, instead shifting gears, "Any others have comments, or? Please chime in about your ideas."

Another colleague then said, "I don't want to be the one who's raising this and speaking all the time. I'm sorry, but one of the questions we did discuss was that cigar [data] is a very good example of where, if you look at Nielsen data, the trends for around the same years, it actually is the opposite of what TTB data is."

He replied with a noncommittal, "Hmm."

Our colleague continued, "So, that raises questions as to the context with which this data is put to use in the assessment and what are the implications of the specific data you use in relation to how it supports decisions related to the application in question. So, that, it raises a broader question. The set of data we use has implications for how it is put to use and how, you know, and that's sort of the tone of discussions we had there, and which is an important point, which is, that's a peer-reviewed article from Nielsen data. There are multiple peer-reviewed articles showing increases in trends in cigar sales, specifically in flavored cigar sales. So, how specific are we getting when we are using TTB data? How broad is it in relation to what the type of product we are trying to use it for? All those are points to further consider."

Our Division Associate Director replied, "Um hm. That's great. Great points. Uh over the same time, is it, I mean, you have different time periods, I imagine, because there's no, you know, I knowing, I'm a person who has dug through a lot of data in the past for like um like certain studies and knowing that this is the quality of the information's just going to vary and because it does, it's not the same source, um. So, then I think it's putting these things on paper and figuring out what seems to be not just credible but most relevant um to say. I think someone had a comment or in the chat. Yeah Christy, 'CTP cites Euromonitor International trend data that contradicts TTB data.'"

We kept trying to explain the situation to him, which wasn't complicated. So, like our Associate Ombudsman had suggested, I'd typed a simple point in the chat, "CTP cites Euromonitor International trend data that contradicts TTB data." Industry data wasn't saying the same thing as other data sources, and we were choosing to use industry data that showed stable or decreased sales trends, so it supported findings of no significant impact. Other data said the opposite.

He continued, "Okay, and, yeah, this is the kind of thing definitely looking for is to see where we are on the map of data and it's, you know, you have to start somewhere and I think just by us understanding where we're at is very critical. It's probably been some time since we've looked at these things, taking a hard look at them but obviously this is the kind of thing too, that uh you know impacts potentially impacts other teams in the entire branch so it would be something to further discuss but we can continue looking at this information so the point here was definitely need everyone's participation in um you know, your opinion, as scientists uh as to what you feel that doesn't have to be like on the spot but want you to be involved in contributing uh what your either recommendation is or what you think needs further analysis. I think that that's a smart thing to do and um but yeah, I like the idea you put a spreadsheet together. That's a good way to track it um and some of the comments. So, yeah, anyone else?"

When another colleague tried to describe the same issue, he changed the subject again and continued in the same fashion. He simply wouldn't engage. Despite his sentiments about being "smart," and our interest in contributing, analyzing, and recommending, subsequent attempts to do that were shut down.

Some leadership regularly prevented productive discussions with these sorts of behaviors. When they did it in writing, I wondered if it was meant to create the perception of a satisfactory response for busy higher-ups unfamiliar with our work who might not read or think critically about what was written. Their tactics became clear when those exchanges were documented in emails, back and forth while they diverted, ducked, dodged, and delayed. It was strange to see higher-ups—some very high up—allow that sort of behavior to continue.

In meetings, they'd likewise talk around issues, over others, or without taking a breath ramble on nonsensically in what seemed like efforts to waste time. They'd just talk and talk and talk and talk and talk. If a meeting was 25 minutes long and higher-ups talked for 25 minutes, that left no room for anything to be heard or handled, but the meeting happened. The box was checked, "We had that meeting."

These sorts of exchanges were increasing in frequency, wasted a lot of time, and further confused staff, all while the quality of our work faltered further.

On April 4th, 2023, Supervisor #2 had advised, "Just to give you my way of dealing with [Branch Chief #3 and #5], when she does that, is you notice how I was quiet a lot. I let her talk and talk and talk and talk and talk, and then I talk. I just let her finish her thought. I just let her go. She shouldn't have, but she will still step on you. I can't really apologize for her because I'm not [her], but that's just her style. And there's sometimes times where I'm just like, 'Oh, [Branch Chief #3 and #5]'s just being [Branch Chief #3 and #5].' And then there are days where I find it really funny. And, you know, I find it frustrating, because maybe I just want to get out of the conversation quick. It's like, I just want to get this discussed, get an answer, move on. And you can't do that sometimes."

He said, "She's not letting anybody talk! She keeps interrupting!" He said, "It's pretty normal." He said, "That's the way it goes. I get stepped on all the time. [Branch Chief #4] gets stepped on." He said, "One time, wow, [Branch Chief #1] just scared the hell out of me," when she even got stepped on, "I'd never heard her yell." Branch Chief #1 supervised her then.

Anyway, we finally met on April 24th, 2024. During that meeting, our Division Associate Director continued the same tactics. For example, talking without pause he said, "There are people, Christy, I haven't had time to talk to at all. They have, I have zero meetings with and they have questions, issues about other topics outside the branch, of course and that's because of the agency has always been strapped for its resources in management. It's not easy. I understand, you know I worked on the outside, that in the government things can be a bit frustrating because we want things to go quickly. Communication should always be flowing. So, I totally agree with you. I'm there. I just don't have the time every day to dedicate to everybody's questions. For example, this morning, you know, I have a deadline to make, another one, and that is something that will actually impact the branch so we can get a leader, a permanent leader in place because this is not my permanent job, 'acting,' you know, being a branch chief for the branch and I'm doing another position. There are other leaders to talk to. I'm actually glad I just want to say that you have reached out to [our Division Director] and [our Office of Science Director #2], and I agree with you. We're thinking, and I thank you for being active participant in, like, for example, the greenhouse gas presentation I think it went really well. I have a series of questions to follow up on that to help the workgroup move forward because that's a very important activity and you're engaged with microplastics. When I was a third, you know, outside of the branch, so to speak, I wasn't even aware so much of these things. And so, as I told the whole, this work is for the benefit of everybody, the entire branch, and they all deserve, I agree, they deserve a permanent leader, Team 1 deserves a permanent supervisor. I'm fighting to get that. That's what I do. I spend my time on those issues in recruitment and um I am listening to you and I definitely want um will be following up on the memo. I'm glad that you're pointing out these details that are important. You're doing your job and that's really commendable. You may not hear it because we don't see each other in person. We're not in the same office and we have to set up meetings. Some days I have a meeting from nine o'clock all the way to four and I can't even address things. So, it doesn't all fall upon one person and that's kind of the experience at the FDA and that's also the beauty of working at FDA is that it takes a team to dream, basically we, to get something done in this agency, it takes teamwork and collaboration. And that involves everybody, like you said from supervisors to staff. And for example, getting a PMTA out, I was a TPL and a TPL has to rely upon science policy precedent, the Office of Science Immediate Office, disciplines, RHPM. There's no way one person can, you do have an impact as one person, but to get, let's say, you spot a scientific issue, then to address it, it is a slow moving machine at times, and that's the frustrating part and I think I told you this in our one-on-ones once that, you know, it is a bureaucracy, but it doesn't mean we quit and you're not quitting. I'm really happy that you're you know, doing your very best to address issues and you care a lot about the work. That makes, that's a public health uh servant and that's what we're dedicated to doing, collaboration that I'm not accusing you at all or blaming you. I would never do that. I would never blame any staff for even a mistake because we're human if they make a typo and put the wrong product name, you know, of course, we want better and higher quality, but they've just proven they're human. And those things can be corrected and we have, you know, we support and we have to make things better, try to look for better references like the project that we had started with looking for better quality information. And I will make time to come into Team 1 to see where things are at and to talk about it and see because I'm the one that started the assignment and I've spoken to [Supervisor #4] several times and what we have to practice is patience and I know that's going to be hard."

I'd heard other staff were counseled for interrupting, so I was afraid to break in, but I did. I tried to cram a lot in without pausing to prevent being interrupted. It was important to accomplish something that meeting was meant to accomplish.

I said, "[Division Associate Director], can I say two things? We're running short on time. I'm sorry to interrupt you but we are running short on time and I really would like to try to make the best use of time and I don't think this is necessarily been a terrible meeting because I feel like you're understanding some things and I'm underst—, I appreciate the workload and I'm not saying this, I feel like I need to say this more because we don't know each other. I'm 53 and I've worked professionally for a long time in different places. I've seen problems. I've seen places that are busy, so I completely understand. I completely understand when things move slow. It takes time. I get that completely. I have a few quick questions that if, would actually really help if we could get to that. I'd like to ask, and we don't need to go on. I mean, I just I want to use, the best use of everybody's time."

At the beginning of the call, I'd reiterated the reason for our meeting. And he'd adamantly indicated clarity in his memories about issues associated with the deceptive approach. I said our Office of Science Director #2 directed me to him.

But he questioned me, "Okay, so you, [our Office of Science Director #2] directed you?"

He knew this before we met. I'd indicated that a few times, pulled up an email that he'd seen that said that, and said, "That is what I've said and what I've written in emails to you. I'm looking at one right now from March 28th, '[Our Office of Science Director #2] suggested meeting with you.' And I had previously said he asked me to meet with you and then follow up with him."

"Okay," he responded, "I'll look back and I'll check with [our Office of Science Director #2] as well. I also have a supervisor. It's [our Division Director] uh that I have to report to and so I'll also check with him as well, just because that's just a normal way of communicating for me um to keep our supervisors informed. I didn't receive any email from [our Office of Science Director #2]."

I knew who his supervisor was and he'd said during this same call that communicating otherwise was okay, even good. And he'd been involved with that email traffic. The emails had gone back and forth for over a month, so there'd been many opportunities to request clarification. So, I said, "I indicated that several times in my email to you, and you could have reached out to me to ask questions. And you could have asked [our Office of Science Director #2] as well."

And his supervisor was informed.

Regardless, he continued, "Oh, okay, uh you know, thanks for the background. Yeah, my understanding, actually I didn't understand anything from the email that [our Office of Science Director #2] um was recommending I meet with you."

I again clarified, "[Division Associate Director], it literally says in the email to you, it says '[our Office of Science Director #2] suggested meeting with you' and we had a series of back and forths where you responded."

I tried to get confirmation about our path forward with respect to use of the deceptive approach. I clearly framed the issue, "We had previously indicated in our assessments that data associated with 32 products could inform an assessment and a FONSI decision associated with 16, and this is by definition a deceptive argument. When this was learned, [our Office of Science Director #2] created a fix immediately, and that fix has been implemented. Since then, and this is something I was going to mention today, I just wanted to get a confirmation on. Since then, [Division Associate Director], you said you agree with the previous approach. And this is one thing that, and this is in an email, and I—"

He interrupted, "No. I don't understand your, that statement."

He did.

I said, "It's in an email."

"No, I understand that," he said, "but I, that doesn't mean I understand what you're saying."

I tried to explain again, "The question is—"

But he again interrupted, "What I'm trying to get at, sorry, is what for me to understand it, is what approach are you talking about just because there are two different things."

I tried to explain again, "Sure—"

But he interrupted, "In order for me, these are your emails."

They weren't just my emails.

I tried to explain again, "I can—"

He interrupted again, "It's easy to basically take things out of context."

So, I tried to explain again, "Sure, I can, I want to make this quick."

He said, "We are out of time, I have another meeting."

He knew what that meeting was about, but diverted, ducked, dodged, and delayed. But I tried again, "Instead of arguing the past, let's go to the future. My question is, will we use that approach or a similar approach again, regardless of what anybody said in emails, what we need is a confirmation. Yes or no, we will be using that sort of approach again? We have a solution on the books. Do we need to go back to that approach?"

He said, "Christy, I would say let's define what approach you're talking about."

I tried again to explain, "Representing—"

He interrupted, "I can't answer that right now 'cause I don't fully understand what you mean by which approach."

He understood.

I tried again, "If we have data for 32 products, will we use it to represent 16 in an assessment again? This was six months—"

We'd worked on the issue for about six months.

But he interrupted again, "Will we use what?! Will we use what?!"

I couldn't get a complete statement out, "Data for 32 products—"

He interrupted, "Are you talking about marketing estimates?"

I said "Yes."

He knew this.

"Hm?" he rudely questioned.

I kept trying, "So, it can be anything. Will we use data that doesn't apply to the products in question? Will we use data that does not apply to the products in question, that can't inform the FONSI? That is a simple question."

But he said, "I can't answer that question because I don't fully understand the context. I understand what you're stating, but I don't understand the context at this point. I definitely want to understand it."

He knew our Office of Science Director #2 wanted to know how our meeting went, so I said, "I'm going to have to go back to [him] and explain that. I'm going to have to go back to [him] and say—"

He interrupted, "Please do and [our Division Director] because, and I will also inform him. I don't I really understand. I do understand that more clarity is needed and I do agree with you on that."

I said, "There is a solution on the books already. There's a solution that's already in the record. The solution that's already on the record—we aren't certain? You were involved with all of this."

He said, "Are you, Christy—I'm sorry. Are you referring to—what meeting? Are you talking about the August one with [our Office of Science Director #2] and [our Division Director] and I where we went over the issue of marketing estimates from there forward?"

"Yes," I confirmed.

He said, "Okay that's what I really need to, basically, we're over time and I understand this is going take a lot more time to go over." He oversaw and was involved with the process and the solution, and the issue was not complicated. It was simple. CTP was pretending data combined for 32 products informed 16 regulatory decisions about 16 individual products. He continued, "I'm sorry, Christy, I'm not going to agree to anything that I don't understand. I'm sure I wouldn't expect anyone else to. I don't understand."

He'd understood earlier in the call and had written about it again as recently as April 5th, 2024, "I support the center's determination and recommendation" in an email where he attached the inaccurate SDR Determination.

So, I said, "You already did in an email three weeks ago. That's the problem."

He dismissively laughed, "Then why are you asking me now? That doesn't make sense!"

I again clarified, "Because [our Office of Science Director #2] asked me to get clarification in a meeting so we don't have to do it back and forth in writing. Because it reversed a decision that took months to put in place. Because you said in writing in an email that you support that [deceptive] approach. And it wasn't even a topic of discussion." He'd brought it up out of the blue in that April 5th, 2024, email. I continued, "I'm sorry, [Division Associate Director], this is—"

He interrupted, "Sorry, Christy. I have to go. We're over time. I have another meeting to prepare for that's on a totally different subject, but it's equally important and I will follow up on this and I appreciate you coming forward. Please, if you're not satisfied, let's collaborate, work together and I believe you're attempting to do that and I do not blame you for any mistakes or for following a process that's erroneous and I'm here to support. It takes a lot of people to make things move forward into work. It's not one person, that's for sure. And so I appreciate [Branch Chief #3 and #5] and [Supervisor #4] for joining and I call for the meeting, and we can adjourn now. Thank you. Have a great day. Bye now."

What? Holy smokes!

He'd invited two others who'd attended, my current Supervisor #4 and our Branch Chief #3 and #5. Interestingly, they didn't bother clarifying. They understood. Our Branch Chief #3 and #5 directed us to use the deceptive approach. She also reviewed and signed documents where it was used. Supervisor #4 had been a primary reviewer back then and used the deceptive approach in his assessment, he said because he didn't want to make trouble for his Team 3 Supervisor #2 who'd just stepped into that temporary position.

All these people knew all of this.

Another point of that April 24th, 2024, meeting was to discuss directives, by leadership that our Division Associate Director supervised, that directed staff to do things that countered our oath and training. This was an uncomfortable thing to do, and I'd wanted to talk with him about what they were directing us to do. But he invited them. Branch Chief #3 and #5 was complicit in it all. My Supervisor #4 had followed those directives and had begun adopting the same sorts of behaviors. They were both complicit. When he invited them, I'd requested a meeting without them.

He'd written on March 27th, 2024, that he could meet on April 24th, 2024, with me and them for 25 minutes. "Based on," he wrote, "workload priorities, that is the time I can spend."

In response I wrote, "Removing [them] from the invite and meeting at the time you set aside on April 24th will add no additional obligation to your workload. I'll go ahead and accept your invite assuming we can do that. [Our Office of Science Director #2] suggested meeting with you. I wanted to follow up about unresolved critical work-related issues."

I'd agreed to meet if they wouldn't be present, but he ignored that request. He didn't respond indicating they would be present. My Supervisor #4 had even told me that he would not attend. I therefore said I was thrown when we started the meeting with them both there.

In the email chain that prefaced that April 24th, 2024, meeting, I'd clarified several times what that meeting was about. For example, I'd written to him on March 28th, 2024, "Some management refuse to engage. That is one thing we needed to discuss."

I wrote that we would talk about "reviewers falsely blamed for following management directives, conflicting guidance from branch management, our current assessment and FONSI preparation and evaluation approach (that lacks guidance and consensus), considerations of marketing estimates, and the directive to consider in no way thresholds as part of our scientific assessment."

The latter was another important topic that had fallen off the table. I'd continued, "Thresholds are implied in any instance when a number or the like is included in the assessment, even if that number is redacted."

In that same email, I wrote, "You are my second-line supervisor, and my first-line supervisor is acting for just a few more weeks. I'd requested a meeting alone and you showed no interest in knowing why before declining my second scheduling attempt, not even after I reiterated the need." He'd repeatedly declined meeting requests without engaging.

I also wrote in that email, "It is not part of my job to pretend this isn't happening or to stop raising important issues. When we meet, I'll be sure to focus on those that are the most critical to our mission."

He avoided discussing mission-critical issues. However, when I asked him if preparers of assessments needed to agree with the arguments in those assessments, he said those arguments needed to be our arguments.

I reported back to our Office of Science Director #2, "The meeting with [our Division Associate Director] today did not go smoothly, but it might not have been completely unproductive. We can talk earlier if you'd like, or I can see how a few things play out first."

The meeting hadn't gone well. But we'd learned the arguments in our assessments were to be our arguments and our Division Associate Director said he'd follow up and that we'd work together. Considering everything, that felt like progress and my job was to work cooperatively and follow the process. I was required to engage and facilitate. So, I followed up:

Hi [Division Associate Director],

Thank you for meeting today. I look forward to working together to address our most pressing needs. Here's one quick follow-up...

The discussion we had about use of the previous approach was in reference to your comment in the email directly to follow [where he wrote that he supported the inaccurate SDR Determination].

What was requested at the meeting today was confirmation that the new approach would be used instead of the previous approach. An example of the new approach is attached, and I'm happy to provide any clarification needed. I can quickly walk you through these documents, but in brief, the first document is an applicant-provided assessment and is followed by reviews with requests for product-specific information, instead of developing a means to use data provided that does not inform the assessment or subsequent FONSI decision. The FDA Office of Scientific Integrity previously indicated concerns about use of the previous approach were not ripe for consideration because this new approach had been implemented.

Additionally, I've been tasked to present this case study and need to understand the outcome. A good solution was quickly implemented that I believe can be framed in a positive way.

Thanks again,
Christy

He never did follow up.

19 | Capitulating

BY THAT TIME, our still newish CTP Director #2 and Office of Science Director #2 might've found themselves at a crossroads. They'd come on board energetic and seemingly sincere and enthusiastic to protect our public health. They'd each said they had an open-door policy and wanted to learn from staff how we could improve. However, their enthusiasm waned.

Staff alerted them about instances where regulatory decisions had been made before reviews and assessments had been completed, and about other concerns. I frequently interacted with our Office of Science Director #2, who encouraged us to continue raising concerns. I'd occasionally ask if I should keep doing that. I was uncomfortable articulating the same regulatory issues to my leadership below him so many times. However, he always encouraged me and us to continue. That was my job and he said we should follow our training.

Over time, however, he seemed frustrated and overwhelmed with his inability to implement change. In an all-hands Office of Science Meeting, he said he'd had his "hand slapped" when trying to make improvements, and it appeared he didn't have control over the staff he oversaw to the extent that he was able to direct work under his purview. Where he'd previously asked to be involved, on May 16th, 2024, he'd now written, "I'm too far from the details to be able to provide direction myself. I have to rely on my deputies and division directors to guide certain aspects of the work."

And certainly, micromanaging every division he oversaw would be difficult. But he'd asked to be looped in, so he now knew about problems. They both knew about problems that shouldn't have been ignored and were deciding how they'd handle that in their new leadership roles. In most cases, they'd not need to provide that direction themselves. Instead, they could've ensured those that they directed, likewise, followed their training.

Instead, they both dropped internal messaging about transparency and their efforts to improve our workplace culture indefinitely went on pause.

20 | Get the Lead Out

A VARIETY OF HARMFUL CONTAMINANTS are emitted into the air when tobacco products are used. Users inhaling from a product experience mainstream emissions. Users, and those who don't use tobacco products, are also exposed to harmful contaminants in secondhand and thirdhand emissions. Secondhand emissions include mainstream emissions that are exhaled by users as well as sidestream emissions. Sidestream emissions constitute the emissions coming off the product that weren't inhaled during use, such as off the end of a burning cigarette or from the bowl of a pipe holding loose tobacco that is burned. Secondhand emissions deposited on any surface are referred to as thirdhand emissions. Thirdhand emissions can also be emitted back into indoor air, creating different harmful exposures in addition to those coming from secondhand emissions. Together, secondhand and thirdhand emissions are sometimes referred to as environmental tobacco smoke or environmental tobacco emissions.

These contaminants accumulate on anything with an exposed surface where a product is used, like furniture, walls, windows, carpets, bedding, clothes, toys, human hair and pet fur, skin, and even food like apples in a bowl on a kitchen counter. Subsequent contact with those items therefore serves as another route of exposure to harmful contaminants. For instance, not unlike concerns with exposure to lead paint chips, a child can be exposed when she puts a toy contaminated via emissions from tobacco products in her mouth or licks her hands after crawling on a floor contaminated via emissions from tobacco products. Harms from these sorts of exposures are one reason the paint industry removed lead from paint products. Likewise, the petroleum industry removed lead from products because of harms associated with lead emissions.

The Center for Tobacco Products was by no means unaware of these contaminants and concerns, and our assessments were to capture what the agency knew during decision making. The scientific community and public at-large had known for decades that emissions from tobacco smoke were hazardous. Consider indoor smoking bans that mean many Americans have never experienced smoking in a hospital, school, or restaurant.

But each year we learn more. For instance, CTP staff regularly attended talks about new concerns, especially associated with secondhand and thirdhand emissions. And CTP had already done its homework, for example, completing in 2018 a 117-page white paper summarizing a three-year effort to identify harmful effects and critical data gaps associated with air emissions.

That white paper was produced by CTP's Environmental Tobacco Exposure (ETE) Workgroup. One objective of that workgroup was to review any evidence of health effects from exposures in areas with unrestricted use like at home or in cars, or at industry farms or manufacturing facilities. The workgroup would identify research gaps and propose projects and policies to fill those gaps. They'd initiate interagency collaborations to address those gaps that had been identified, and other related needs. That group explicitly focused on concerns about chemical residues left behind, in indoor and outdoor environments, after tobacco products had been used there.

In that 117-page white paper, all six involved scientific disciplines communicated significant warnings about secondhand and thirdhand contaminant exposures from tobacco products. Known harmful effects were communicated. Alarms were repeatedly raised about risks to the unborn, infants, and youth from well-known harmful effects and owing to the breadth of what is not known. As important as what is known, particularly from a regulatory health perspective, is what is not and how to handle that. Warnings about how lacking information across all tobacco product categories precluded informed regulatory decision making, that therefore risked the public health, were communicated from the Office of Science's Chemistry Branches, Engineering Branches, Epidemiology Branches, Medical Branches, Toxicology Branches, and from the Environmental Science Branch.

Lead is just one harmful contaminant identified via that 2018 effort. Many tobacco products release lead via mainstream, secondhand, and thirdhand emissions, and lead causes brain damage.

Also part of the U.S. Department of Health and Human Services, the Centers for Disease Control and Prevention (CDC) warned that lead's damaging effects can be lifelong and permanent and include speech and hearing problems, learning and behavior problems, slowed growth and development, and damage to the nervous system and brain.

Online, the CDC provided for the public a long list of consumer products and activities that can expose us to lead. They listed paint, soil, water from lead pipes, toys, jewelry, candies, and cosmetics. They listed traditional medicines, aviation gas used in piston engine aircrafts, moonshine, and the fabrication of artistic products like leaded glazes and stain glass windows. They mentioned welding, casting, smoldering, mining, smelting, sanding, incineration, demolition, recycling, and remodeling. They indicated lead can be in vintage items such as leaded crystal pieces and dishware, ceramics and silverware, painted tin panels, and furniture. They mentioned the manufacturing of metal equipment parts, batteries, circuits, and bullets. They even mentioned lead exposure from retained bullets in gunshot wounds, a source of lead exposure unfortunately on the rise in the U.S.

Significant efforts have been made to prevent exposure to lead, like removing it from paints and gasoline, and from some ammunition. Workplace protections and other warnings help prevent exposures. And Americans were reminded again about its hazards, especially to children, during the Flint River, Michigan, water crisis.

In 2024, the FDA removed from the market applesauce with cinnamon that was contaminated with lead. The FDA specifically cited concerns about children that might be exposed to lead and the agency's role protecting the public health in its press releases about the recalled applesauce. Further, the CDC stressed the importance of and provided guidelines and resources for prevention and testing—indicating there is no safe exposure to lead.

There's no threshold below which exposure is safe.

However, with respect to tobacco products, the FDA neglected to inform the public about what was known to the agency about lead exposures. That was the purpose of our assessments, to inform the public. The CDC, where our CTP Director #2 worked for about a decade on related issues, indicated the most common source of lead exposure in young children is dust coating their hands and other objects that they put in their mouths. So, our children are exposed via secondhand emissions and eat lead deposited as thirdhand emissions in their homes and in cars and in any other spaces where those products are used.

That said, emissions from tobacco products contain a lot more than just lead.

21 | A Lesson in a Song: "You Turn the Screws"

IN EARLY 2024, our Division Associate Director asked me and six other Team 1 scientists to update the section of our assessments that informed the agency's decision making about air quality. We started by updating the references. We summarized from those references information about the sorts of contaminants released into the air when tobacco products are used. We compiled information about the prevalence of exposure and the harmful effects of those contaminants. We compiled information about the levels of those contaminants, or exposure concentrations, anticipated to cause harmful effects. And we described the groups of individuals vulnerable to those exposures. We began updating the narrative and presented some of that information in a concise table that would be included in our assessments. We'd started with cigars but, fortuitously, that search returned references that facilitated critical updates about other tobacco products.

February 5th, 2024

We received the assignment during a meeting with our Division Associate Director, just after he'd also become our Team 1 Supervisor #3. He said, "I was looking at the literature that's used to support some of the statements in our assessments. I'm thinking it could be a good idea to revisit it to see if it's the most up-to-date information. And I guess it would be sort of my job here to point out what those are, and then I'm thinking, basically, let's look at the literature, this could be related to environmental justice, I think if I remember right, or secondhand—actually, I think secondhand smoke might have been one because we want our assessments to contain the most up-to-date information and the best available information."

Then at that same meeting, instead of using the most up-to-date and best available information, he said we should keep the information within the boundaries of the statements in the original assessments, despite any science that might indicate otherwise. He said, "Perhaps we could look at those papers and see how outdated they are, and then you know, make an assessment and then develop, you know, a recommendation as to potential papers that could substitute to support the [existing] statements."

His directive to ignore current science and only select newer references that support previous claims was synonymous with us reinforcing in 2024 the FDA's 1983 claims that previous formulations of ranitidine, used to treat gastrointestinal conditions, were safe using 2004 data. We've since learned about its contamination with the carcinogen N-nitrosodimethylamine. Imagine, when we now know more, how damaging dated science used to support previous agency claims in current decision making could extend. His directive unequivocally contradicted our oath and training. As regulatory scientists, our role was not to find information that supported a certain claim but to objectively consider the most up-to-date, best available information.

February 14th, 2024

Then during a meeting with only the two of us, he said, "The reference project activity that I opened up on air quality, just to see if the quality of the references, is to start to visit the question of whether we have the best available information that we're using, to look at the papers and see if there's anything new that's credible and then, but not to jump to a conclusion that, 'Yes we need to just switch this reference,' but let's talk about it and analyze the information if there is."

That sounded better.

He continued, "Let's start the conversation and see if you can find information that you might think could be improved upon what we have, and then we'll talk about it, as to why."

February 20th, 2024

When asked during a Team 1 meeting, "What's the timeline on this project?" he said, "There is none. There's not a deadline."

He explained, "The timeline—it's really most important to be careful with what we're doing, it's driven by that. I'll put it that way. And so that's what makes it better, because there's not someone breathing down your neck, at least right now, saying you need to do this by next Wednesday, which is a typical thing to be honest, at least on the Tox [Toxicology] side. There's a lot of tired folks over there, certain teams that have been working on memoranda, but for good reason."

It was a shame to hear other staff working on memoranda to clarify decision making in our division might not have time to do a good job.

He continued, "And so I'm like a type, I like to be kind of, to be preventive, to, you know, like, work on things, to assess the, the information."

It was good that we'd have time to do a good job, particularly considering the breadth of information about contaminants emitted during tobacco product use. Secondhand emissions from tobacco products kill tens of thousands of Americans annually and cause harmful, sublethal effects to many more. As the U.S. public health agency that regulated tobacco products, we needed to be sure our work reflected the most up-to-date and best available information. We needed to communicate the most critical risks.

March 7th, 2024

During our next meeting, the overall purview of our branch had been captured via an exchange between our newly appointed Team 1 Supervisor #4 and a colleague who'd recently come on board. We'd just talked about evidence that more individuals were exposed to secondhand emissions than previously believed. That was something we'd need to update in our assessments.

The colleague asked, "So, the thing I keep circling back to with this is, 'Why are we doing this?' Because we're still going to authorize it [the tobacco product]. You know, I know secondhand smoke is bad and regardless of the trend, whatever it's doing, we're not going to say, 'Well, because of secondhand smoke we're not authorizing you to sell this.' So, it seems like this is just a bunch of us talking." He said, "What are we doing here to put all this effort into something that, like, eh, you know what I mean? So, help me battle that in my brain, because I understand we got to do it, but it just, it seems very silly."

Our Supervisor #4 explained, "Yeah, I get that. But, I mean, so NEPA is about information sharing and so, I mean, if you look at a lot of assessments, it's not saying whether or not something can go ahead, it is informing the public of potential issues or, you know, environmental impacts of that action. And so for that, like these assessments, we just, we want the most up-to-date information, so people are aware."

With respect to our assignment, Supervisor #4 clarified, "The purpose of updating information in the assessment, the point is to have the most up-to-date information on impacts that's available."

Our Division Associate Director had now also become our new Branch Chief #2, and Supervisor #4 who reported to him had taken over Team 1. Supervisor #4 had been introduced as our new supervisor at the February 20th,

2024, meeting where the assignment had been discussed. Therefore, he heard our outgoing Supervisor #3 and Division Associate Director explain, "It's really most important to be careful with what we're doing, it's driven by that, I'll put it that way. And so that's what makes it better."

He'd heard him say, "And so I'm like a type, I like to be kind of, to be preventive, to, you know, like work on things, to assess the, the information."

At that March 7th, 2024, meeting, a colleague said, "We could expound for pages," because there are so many known and potential impacts.

Supervisor #4 said, "When we come to an assessment, we want the best available data in that as a source of information for people to know what types of impacts there are." He said, "It seems like there's a lot of interest in secondhand smoke and a lot of science available." He said about some parts of our assessments that excluded important known impacts, "I don't know why," and said, "I will have these conversations with leadership because I share the same questions and I want this, I want us all to be involved in this process."

He said, "Don't worry about it. Bring it up. Don't worry about me. I'll take it. I'm here to plop my desk chair down on that grenade, so let's talk about it."

March 18th, 2024

During our next meeting, we talked about how we could quantify by the unit, like a single cigar or a single cigarette, the amounts of contaminants emitted into the air. We talked about how we could calculate those amounts in micrograms, but in some cases, emissions associated with a marketed product were so great we'd need to convert to kilograms. We talked about references that would help us consider mainstream, secondhand, and thirdhand emissions for any product that emits emissions. So, we'd also be able to apply the same approach to ENDS products. We talked about how that was especially important indoors where contaminants accumulate and constitute a constant source of harmful contaminants, even long after the use of those products had ceased. We talked about our ability to do that for a variety of harmful contaminants.

We talked in detail about some specific emissions, for example, about the amount of lead emitted and our ability to help the public understand exposures that the agency's decisions to grant marketing orders could subject them to. We were told to, as best we could without unnecessarily dumbing things down, use clear language and examples that were easily understood by the public. We talked about providing that information per pack, one colleague explained, "When people smoke, they know when they smoke a pack of cigarettes. So, it's like, okay, 'When you smoke 20 cigarettes, this is how much lead you are emitting into your house where your child is.'" We could do the same thing for single cigars and multiunit packs. We could likewise apply the same approach to ENDS.

The amounts emitted and that accumulate are not inconsequential.

We talked about the amount of lead emitted from one cigarette. And while it's unlikely that one child would take in all of that lead from one cigarette, we talked about exposure to "a child in a house or a car with an individual who smokes [and that] most people who smoke probably smoke more than one cigarette a day." One colleague suggested we provide information about contaminants emitted over the course of a year "that says, 'Hey, look, you're gonna have this much lead in your house if you smoke a pack a day.'"

Supervisor #4 said, "Yeah, I definitely, I agree that the pack-a-day is like the most relevant, that's the most easily interpretable metric for use of that."

He said, "Yeah, there are studies that document disproportionate concentrations of lead in blood concentrations in kids."

We were objective and framed what we could communicate to the public in appropriate contexts, for example, considering uncertainty and understanding we'd need to clarify that contaminant concentrations would vary, and that there are all sorts of places these emissions could end up as well as mechanisms to remove emissions from indoor air.

One team member said, "We know this information exists, it is available and we can't really ignore it, so we need to figure out what to do with it. And lead is just one contaminant. We have mercury. There is a huge list of these different contaminants." So, we talked about that.

Supervisor #4 said, "Looking at this data, it starts to make more sense how these things are affecting people in the places that they live. And I think that's really important to consider."

He said, "We can think about ways of presenting that that might demonstrate the potential concerns but not reveal any confidential market information. There's definitely a way to do that."

At the end of the call, he said "Keep working on this." He said, "Keep chugging away on the revisions."

March 21st, 2024

We'd begun moving some information into the example assessment from a spreadsheet where we compiled new information. And while our directive, as I heard it, had been consistent for a few weeks, other staff reported contradictory messaging from some leadership indicating we shouldn't update our assessments to reflect the most up-to-date and best available information.

So, when Supervisor #4 asked for an update, a colleague said, "What the discussion recently has been is that there's a difference between updating sources for existing text and existing topics, looking at the recent literature, and thinking about if there's any additional subjects that need to be added to that section, right? So, if there's any new science that has come out, on secondhand and thirdhand smoke, that should probably go in there, that isn't already topics that are touched on in there. And that's kind of what the recent discussion has been right now, right? I was hunting updated references for the existing language, but there may be new language that needs, new entire language into subjects that need touched on based on the state of the science now, right? Especially, I think, thirdhand smoke would be one and, you know, new information we've learned about, more specifics about how this stuff affects people."

Supervisor #4 responded, "Okay, cool."

He encouraged us to add that additional text, that we'd been compiling in a spreadsheet, to the assessment that would serve as an example of what would be communicated to the public. "Yeah," he said, "I looked through, I was looking at the spreadsheet and the example assessment, and I noticed that some people started adding text to the example assessment, but that there was a lot more information in the spreadsheet. I encourage you to start making the edits in the example assessment."

A colleague asked for clarification, "So, would the draft assessment include," not just information in the old version of the assessment, but also, "information that we think is also relevant to be included?"

Supervisor #4 confirmed, "Yeah, I think that, you know, there should be, if you find updated information and you have to revise the text to correspond to that, that's what we're shooting for at this point." In respect to our discussion focused on lead, he said, "And there's literature that shows that there's, that certain demographics have higher lead exposure that's correlated to cotinine exposure. So, I mean, that's pretty good evidence that it's coming from secondhand smoke."

The common ingredient in all CTP-regulated products, except smoking accessories like rolling papers, is nicotine. While nicotine is not credited with many of the negative health outcomes associated with tobacco products, it is the addictive ingredient that makes it hard for users to stop using those products. And that makes it harder for addicted users to avoid those negative health outcomes. However, some of nicotine's other characteristics provide insight into how those products are used and the potential breadth of their effects.

Anything we take into our bodies is subject to potential changes, meaning that what we take in can be converted into something else. Such *biotransformation* includes harmful and helpful mechanisms, for example, that break complex nutrients down into compounds that fuel our cells and build our muscles. Those changes can also aid in excretion, for example, via urine. One product of the biotransformation of nicotine is cotinine. Among the many other biotransformation products of nicotine, cotinine is a common one that persists for a bit in the body and that is fairly easy to detect and quantify.

Very, very low levels of cotinine might be an outcome of eating some vegetables that contain very low levels of nicotine. However, higher levels of cotinine detected in biological samples like blood and urine are used to monitor exposure to products that contain nicotine. The amount of cotinine in a biological sample is related to the amount of nicotine in a product used, and the amount of nicotine is proportional to the product's other constituents, like lead. Therefore, cotinine can be used to estimate exposures to contaminants emitted into indoor air from tobacco product use, like lead.

This is the sort of thing Supervisor #4 said constituted evidence of lead exposure from secondhand emissions that we could include in our assessments. We'd of course need to include any uncertainties or caveats associated with those correlations. But that is something we could've done to help inform those demographics and the public, overall. He said, "There's potentially holes that we need to fill based on more recent literature. And so, I think we can do that. Yeah." He said, "Part of [our Division Associate Director's] original assignment was, you'll come across information as a group, you'll have to discuss it and come to consensus about whether or not we should include it."

March 28th, 2024

At our next meeting, Supervisor #4 started shifting gears. He struggled from then on, vacillating between credible work and the status quo. Over time, these sorts of behaviors revealed how maintaining questionable practices as a means to hide them, because changes would reveal what had come before, is fraught with trapdoors and will eventually fail. Credible policies could be articulated broadly in writing, but the status quo couldn't and therefore required significant unwritten micromanagement via word of mouth.

I'd come on board in February of 2021, and from the onset things had been disorganized. There'd been changes to staff in our division. And it was a year before the Biden administration appointed an FDA Commissioner in 2022.

Shortly to follow, our CTP Director #1, Office of Science Director #1, Office of Management Director #1, Branch Chief #1, Team 3 Supervisor #1, and others left CTP. Remaining leadership jumped in and out of various positions while our division and branch doubled in size. So, the number of siloed conversations needed for complicit leadership to coerce predetermined findings with unconvincing arguments became hard to keep up with. We worked remotely, so pinning us up against the bathroom door with our pants down would be difficult.

Dwindling, long-term leadership fluent in the current thinking assumed an ever-increasing burden. For them, maintaining the status quo meant they micromanaged downwards, sideways, and upwards, especially when their new leadership wasn't yet aware that our work wasn't up to snuff.

Putting those lacking experience with CTP's status quo in leadership positions complicated things. Our CTP Director #2 and Office of Science Director #2 recently came from outside FDA. They touted our credibility and asked staff to raise any concerns, so some did. Open positions involved with our assessment process were subsequently filled only by internal hires already complicit in questionable practices.

Indiscretions were seldom or ever truly resolved. Some higher-ups avoided addressing problems, some that possibly went unnoticed under their purview. Some might've drawn litigation if handled properly. Instead, the status quo was maintained explicitly to avoid calling into question agency credibility that would undermine its broader role protecting the public health. For example, some said we needed to keep CTP's problems quiet as to not undermine FDA's important work with vaccines. Compelling, but gross. To prevent measles, we'd need to facilitate lung cancer.

Once higher-ups decided to hide those problems and allow questionable practices to continue, they'd need to learn them so they could also enforce them. However, they often didn't have the time or expertise, so their involvement muddled things further. And remember, articulating the status quo in writing was a no-no. They'd therefore be beholden to complicit staff they supervised.

All these cooks in the kitchen working without a recipe made a mess. And many of those individuals who maintained the status quo sat in important roles where we needed them to focus on our mission, but they were frazzled. Keeping all the lies straight required a lot of energy. They were busy putting out fires in lieu of paying attention to the science and to CTP's public health mission.

Anyway, Supervisor #4 began that March 28th, 2024, meeting with, "So, I do want to say with this assignment, like the last few weeks we've really expanded.

And I think that's like, I like that part of science is that, like when you start discussing a topic it's like the big bang and it kind of like expands at light speed. And I think that's fun."

Fun? This wasn't grad school. Where was this going?

He continued, "And so I think perhaps because of that I probably let it go a little farther than the original assignment. I think that's good. So, but now I think that there's edits on the document, and so we can start reigning it in and discussing what to include or what to exclude. And I think that's going to be a fun conversation."

He said he let it expand because it was "fun." The assignment was to carefully analyze and assess the most up-to-date and best available information about deadly and otherwise harmful emissions. He'd previously, repeatedly, provided that same directive, for example, he'd said we were "informing the public of potential issues … and so … we want the most up-to-date information, so people are aware." Until that information had been identified, we'd need to identify it, regardless of if that was fun. Because the assignment came from an individual several levels above him in the hierarchy, this change was concerning.

Supervisor #4 continued, "I've talked to [our Division Associate Director] since then, and I think that maybe he had some ideas, but his, what he told me was that he really wanted to focus on the discussion of the science and the, like, consensus building about what information to include."

That's what we'd been doing.

He continued, "I also think that, like, it expanded really fast to a point that like, was never-ending."

It wasn't never-ending. Secondhand exposure is a big topic. It had been just a few weeks, and we'd worked on this part-time while prioritizing other work. But we knew how to focus and had begun identifying the most critical concerns. We'd made good progress. He'd seen that and said that, and this was weird.

One colleague said he was concerned that discussing what to include or exclude before we'd identified the most up-to-date and best available information seemed like cherry-picking. But it wasn't even that. Our basket of cherries was, at best, half full. What Supervisor #4 was moving us toward constituted unscientific and uninformed voting and was not consensus building.

"So, I think that, like, we have to compartmentalize," Supervisor #4 said, "and so, we're looking at this one, at the air quality of the cigars, trying to update the information that's there, that, as we've talked about, like if you update references, there has to be associated text that's updated. With that in mind, I wasn't thinking about like a complete restructuring or a rebuild of it." He then begrudgingly said, as if we weren't supposed to edit the assessment, "We'll see what happens, because I think there's been some edits made in the document."

What?

We hadn't gone rogue. Just days earlier, he'd been the one who directed us to move information there from the spreadsheet. And there'd been no discussion about limitations to moving text around, "restructuring" or "rebuilding." What difference would any restructuring or rebuilding make if the objective was to ensure our assessments contained the most up-to-date and best available information?

We tried to adapt on the fly to this strange change. For example, I said we could add information about secondhand emissions. I then sheepishly proposed that we could also easily add information about mainstream and thirdhand emissions, if that was something leadership wanted. He then changed his tune, saying, "Yeah, that's awesome."

What?

We then provided general updates relevant to all combustible products, but to prevent additional criticisms, carefully focused our detailed discussion mostly on cigars and secondhand emissions. We shared conclusions of studies that showed disproportionate negative effects of cigar smoking on low income and minority communities. We said the new references we were tasked to identify also indicated cigars can be more harmful to children than cigarettes, can emit higher amounts of harmful respirable suspended particles and other contaminants, and can potentially impair vascular endothelial function and increase the risk for cardiovascular complications. We also shared conclusions from studies that indicated secondhand emissions may increase the risk of nicotine dependence and the likelihood of taking up smoking.

We talked about increased cigar sales data that therefore could indicate granting marketing orders for cigar products could increase negative effects. This might make it more difficult for the agency to reach findings of no significant impact that relied on decreasing trends, or minimally, we might need to do more work to reach those findings.

We talked about how conflicts of interest might compromise industry-reported data that we used, for instance, that showed decreased cigar sales. We'd asked if we could remove industry-reported data, especially when contradicted by more objective sources. Branch leadership obfuscated efforts to critique data from the industry, despite requirements that we vet it for accuracy and objectivity.

And our CTP Director #2 had recently written in his SDR Determination:

FDA is responsible for the scope and content of assessments (See 21 CFR 25.40(b)). "Generally, FDA requires an applicant to prepare an assessment and make necessary corrections to it. Ultimately, FDA is responsible for the scope and content of assessments and may include additional information in environmental documents when warranted." Furthermore, "The agency evaluates the information contained in an assessment and any public input to determine whether it is accurate and objective, whether the proposed action may significantly affect the quality of the human environment, and whether an environmental impact statement (EIS) or a finding of no significant impact (FONSI) will be prepared."

So, we needed to determine if data we used were "accurate and objective." So, during that March 28th, 2024, discussion, we suggested adding another source or sources alongside a disclosure that trends were disparate. This was an attempt to placate branch leadership. We could add an additional source or sources and say that data sources conveyed different things without pointing fingers.

But we were supposed to determine if information was accurate and objective. We couldn't avoid what we'd learned and talked about how disparate data sources that indicated different things, such as disproportionate negative impacts in urban areas, might be important information that we needed to consider. While we'd learned a lot of new things, what we communicated still fell within the purview we'd been assigned—within the most credible purview.

I mentioned that we could compare mainstream, secondhand, and thirdhand contaminant emissions with concentrations associated with a risk of harm, alongside an explanation of potential exposure, that something present doesn't necessitate uptake and effects. This would help us quantify potential effects framed by uncertainties. And we talked again about improved methods to quantify exposure to secondhand emissions.

We indicated there was quite a bit more, that while our effort didn't yet reflect all the most pertinent up-to-date information and best available information, that we were getting there.

Now, Supervisor #4 said, "This is looking like really awesome information. I'm like, this is a good reveal. This did not let me down. Like, I kind of got the sense that there was activity in the document and like people were working on this and it's awesome. I love to see this, these edits. I'm stoked." He said, "This is awesome. This is fun." He said, "This is really good. I think that my neurons are firing, like this is really interesting. And I'm excited to start the conversation about the science."

We'd been discussing the science, but that was okay.

He said, "The next steps for us, is once we consolidate and agree upon the language of this section, it's basically like proposed changes to the assessment, to our standard assessment, and we'll send that to branch leadership, and they will review it and make a decision on whether to accept it or not." He said, "Yeah, and like, you know, my approach to this is that we're still kind of in the expansion phase where we're adding the relevant information that we find. And there are ways to consolidate."

What?

Okay.

April 4th, 2024

Supervisor #4 started our next meeting with Beyoncé's rendition of Dolly Parton's "Jolene."

I said, "It looks like we have a playlist on the screen." The video and song then played while we tried to talk about work. We wanted to focus on carefully analyzing and assessing information to ensure our assessments contained the most up-to-date and best available information.

I later asked AI if it was appropriate to play that video in the workplace and it said, "No, it's probably not appropriate to play Beyoncé's 'Jolene' video in most workplaces. While the song itself has a strong message of female empowerment and standing up for yourself in a relationship, its lyrics and themes might not be suitable for all office environments."

"Why are we listening to 'Jolene'?" a colleague asked.

"Have you all listened to this album?" Supervisor #4 said, "This is a great cover. I really love this. But I have a lesson in this song. I was really inspired by this album. It's really amazing. Okay, anyway, we'll get back to that. He then said a bunch of positive things about our efforts and mentioned we'd soon hear updates about our work priorities with respect to an ongoing reorganization of CTP's Office of Science.

Our Office of Science Director #2 was involved in that reorg and on February 29th, 2024, had just said, "Some scientists aren't good managers and the habit of burying problems by managers will stop." He framed some needed improvements in light of previous evidence of wrongdoing at CTP and a December 19th, 2022, review of CTP by the Reagan-Udall Foundation for the Food and Drug Administration, "an independent 501(c)(3) organization created by Congress 'to advance the mission of the FDA to modernize medical, veterinary, food, food ingredient, and cosmetic product development, accelerate innovation, and enhance product safety.'"

During the October 24th, 2023, public meeting, "Premarket Applications: Opportunities for Stakeholder Engagement," our Office of Science Director #2 had said that Reagan-Udall Foundation review had raised challenges and opportunities. He said, "Despite numerous challenges posed by the application review, including managing the volume of reviews, weighing complex and evolving scientific evidence, and responding to legal decisions, CTP has made significant progress in reviewing premarket applications and is continually seeking opportunities to enhance efficiency. As [our CTP Director #2] yesterday said, we are committed to operationalizing the Reagan-Udall's recommendations to strengthen the regulatory process."

He said, "We see these recommendations as opportunities to improve efficiency, effectiveness, and transparency in our scientific review process."

That review said, "CTP has been forced to operate primarily in a reactive mode, moving from one challenge to the next, mainly provoked by the outside forces discussed." They discuss that, "Applying new regulations to a large, existing industry is difficult, and CTP's task has been made even more difficult because of the sheer volume of product applications, changes in agency leadership, and near constant litigation it has encountered." They recognized, "That to improve the effectiveness of its application review, the center should make process improvements and identify and address the policy and scientific questions that underpin its regulatory framework."

In that context, our Office of Science Director #2 said, "We are developing a more efficient framework for high-quality tobacco product applications, increasing internal communication to improve scientific engagement and deliberation."

And while he said transparency, internal communications, and our science would improve, he said some weird things. With respect to what applicants should provide to CTP to be granted marketing orders, at the same meeting, he said to industry, "We won't necessarily be saying, 'This is where you need to be,' but the breadcrumbs will be on the trail for you to follow as the body of evidence and body of information increases." He said, "Our job" at CTP "is to look at each individual application and the information provided. Over time, you'll be getting signals and more signals and those signals will guide you one way or the other and which will likely increase your chances of getting a positive order."

"Breadcrumbs will be on the trail"? "Signals and more signals"? Why not just put policies and processes in writing? And was our job to facilitate chances to get a positive order or to do objective regulatory science?

Anyway, because Supervisor #4 was looping us in about upcoming changes and improvements during that April 4th, 2024, meeting, a colleague asked about our priorities. Supervisor #4 responded, "Our priorities were and will be related to information in the Environmental Science review." He then clarified, "We need to justify how [what we do] is relevant to the assessment," framing what we did within the context of our branch's role in CTP. This was messaging we constantly heard. Everything we did needed to be justified within our role implementing the law. That made sense to keep us focused on CTP's regulatory compliance.

"So, okay, so moving on to the assessment," he then said in reference to our ongoing effort to ensure our assessments contained the most up-to-date and best available information. He said, "That was awesome. That looks really good. I really, I liked reading through that. We'll kind of go through my comments. I'm going to kick it back to you all to do the revision because I think that this is, I did feel very powerful while reading through there with the power of the editor, but it's not my words. It's our words. I made some suggestions on places that could be organized and consolidated to be more concise, but I want to send it back to you all to make those decisions. And so, you know, we'll go back and forth. I do, I think this is great."

"Okay, so back to 'Jolene,'" he said. There was a pause as he fumbled with the controls and he said, "This, okay, so Beyoncé put out like some country music, and she knew that there would be criticism of that. And so, what does she do? She covers one of the most recognizable country songs ever in 'Jolene.'" With the video playing and the volume almost too loud to hear him, he continued, "Knowing that a lot of people might have feelings about that, Dolly Parton introduced it on the album. It's genius! 'Cause if Dolly Parton approves of it, who cannot? Right?"

He then connected what Dolly did for Beyoncé to our assignment, "Okay, anyway, a side note, just to say that in writing this [the updated assessment], we have to prioritize information, and *we* decide what is most important and how we justify it. And I think that we can anticipate some of the reviewers' comments and address those in our justification and in our writing. And so, I just wanted to mention that." He then said we could enlist others to advocate for us like he said Dolly did for Beyoncé, "Yeah, Al Gore. Thomas Novotny. Who else is out there that we could have introduce it?"

Where was this need coming from to shield us from grenades while implementing our agency's well-known mantra reflected in an assignment to carefully analyze, assess, and ensure we used the best and most up-to-date information? Was that not what those reviewers of our work—our leadership—would want? Why was this assignment so controversial that it required defending by those known for bravely opposing the entrenched current thinking?

He then said, "Okay, anyway, um, we can talk about Beyoncé later, if anybody wants to."

He then referenced an on-screen version of the example assessment we'd been working on, "Okay, the assessment, I put this note here. There is no word limit. There's no requirement. I just put this as a reference, because when we first started looking at this, [a colleague] looked at the air quality section and it was like, 'This is too long. Why don't we just have one sentence that says there's no safe level of exposure?' And at that point, the air quality section was 600 words with 11 references. The environmental justice section was 24 words with zero references. The version that was sent yesterday was, the air quality section was 1,448 words with about 35 references. I kind of went through and counted, but I was trying not to count things that were cited twice. There were a lot, so I couldn't handle that mentally. And the environmental justice section was 600 words with nine references. So, that is something to keep in mind."

What was it we were needing to keep in mind? We'd increased our treatment of substantiated harms associated with tobacco product emissions from 600 to 1,448 words. The section about environmental justice increased to 600 words from 24 words and zero references. How on earth had a section associated for so long with known harms been limited for so long to only 24 words? And how did that matter? We'd been told to carefully analyze, assess, and ensure our assessments contained the most up-to-date and best available information. Why say there's no word limit and then count words?

We listened quietly as he continued, "So, going through this, I didn't, I made a few like actual revisions to this, but not too many. Like I said, I would like you all to do that. I pointed out things like where we might be able to make things more plain language and concise and accessible. And I think that this process, when I do scientific writing, expands greatly, and I think that the way that this came out is really great because this section has all the examples of different effects."

But it didn't. We'd been clear that we were still carefully analyzing and assessing the information to ensure our assessments contained the most up-to-date and best available information.

It was like they'd say it was complete, so we'd think it was complete.

We were regularly directed to keep things concise, and that made sense. But our work was also required to constitute robust and comprehensive science. Those can coexist. However, some leadership referenced the need to be concise to justify excluding the most up-to-date, best available information. This sort of behavior, to dismiss or preclude considerations of potentially important information or known harms, was becoming more common.

Supervisor #4 wanted to exclude even a mention of contaminants with serious harms, some before we'd talked about them, and some before we'd even had a chance to tell him what they were. It was as if listing just any three examples of not-too-bad effects would do the job, when we knew there was more. Instead of focusing on the spirit of the law aimed to inform the public about known and potential harms from tobacco products, he'd rally the team to fill the space with some text that made it appear we did the job. We wouldn't bother identifying an inclusive list or even a more comprehensive list of effects from which we'd select those that were the most problematic.

Everything about our efforts indicated we understood that the presence of something in the environment doesn't necessarily result in an effect, or even an exposure. Yet, he continued with that additional rationale for excluding information he hadn't yet seen, saying, "Like, because something is present in secondhand smoke and is measurable, that doesn't mean that there's an effect."

That can be confusing. Being in a room with something, for example, seems like exposure. However, exposure happens when there's been some sort of interaction, for example, when a contaminant has entered the body after crossing a biological barrier such as the skin, respiratory tissues, or gastrointestinal tract. Otherwise, you and that contaminant are just in the room together. And even when there's been exposure, that doesn't mean there'll be a harmful effect. Our bodies have mechanisms that alter, compartmentalize, and excrete contaminants, and therefore can reduce, eliminate, or even increase potential harms.

We weren't trying to force unsubstantiated concerns into our assessments. There was no need for that. We knew the difference between that and conjecture, and there were plenty of substantiated concerns. We'd offered up the means to collate, calculate, and concisely share the most critical information about harmful contaminants. We'd indicated there were many, so we'd need to talk about how determinations would be made about what was included and excluded. Our aim was for our assessments to be objective, accurate and appropriately representative, and that we disclose an appropriate level of uncertainty. We knew we needed to work efficiently, so we asked about adding information developed in-house by CTP scientists in other disciplines.

He responded, "Oh, okay. Yeah, let me think about that. First instinct is that is outside of the scope of the project updating references in the assessment and more in the box of scoping. That seems like a big picture topic. But let me think about that. And I might reach out to you and discuss that more." However, there was no follow-up about our use of those resources.

Supervisor #4 then talked about how he'd deleted from the assessment a colleague's mention of harms from mainstream emissions. He said, "I deleted that just because it's kind of, it's um, it's *known*. I just wanted to."

What?

Internally, CTP staff were at odds with each other. When concerns about inadequate efforts to inform the public were raised, some leadership and some colleagues said with tones of judgement and disdain that people should know better because "everyone knows about the harms associated with tobacco products." Those individuals and some CTP staff who used nicotine products seemed uninterested in the credibility of our work products. Some leadership and some colleagues said what we did didn't matter, that CTP's decisions about marketing orders were made regardless of any science.

However, CTP justified the existence of its own outreach program on the premise that many people don't know or understand. Further, CTP justified its involvement in research on the premise that many harms weren't yet known or well understood, and we needed to communicate those uncertainties as well.

And what about the babies breathing those secondhand emissions and rolling around in those thirdhand emissions? Stupid babies should know better. Stupid babies shouldn't be putting things in their mouths or breathing air in homes where others they can't control use tobacco products. They can't read, and certainly their parents and caregivers and people every place they go already know about the science associated with secondhand and thirdhand emissions. Do none of those Americans need the government agency tasked with protecting the public health to take that role seriously instead of engaging in machinations?

If we polled people about their knowledge of thirdhand emissions, would it indicate that everyone knows?

Anyway, the point of the assessments was to inform the public, and the rationale that we leave things out because "it's known" would exclude a lot. So, I said, "We don't assume people know."

Supervisor #4 then backpedaled and said, "We should not assume that people know things, I guess, you know, in reading through that it seemed like it was not directly related to air quality. It is important context. It's a suggested edit. I think that, you know, if you want that, you can reject the edit."

A colleague acknowledged and agreed that assessments should use plain language, "But that doesn't mean we can dumb down on the concerns about these chemicals that we as scientists know. So, maybe someone can be consulted like, 'Okay, we have these concerns based on the literature.'"

Supervisor #4 responded, "I mean, yeah, it's like, it's not always the best approach to delete something to save words. Like, I don't think that makes sense. But, like, you know, we have to in the assessment, we kind of elaborate on things in proportion to their potential impacts. That's the NEPA thing."

We were absolutely trying to "elaborate on things in proportion to their potential impacts," while it seemed he was still counting words without giving a crap what those words said. One rationale used to exclude information was that some people wouldn't understand, and that rationale could be used to systematically exclude everything. So, I said, "I know that there are assessments out there that include data." I said, "The assessment is supposed to be accessible, and it's supposed to inform the public, and the public also includes scientists and people who have educations and can understand things, and the idea is that you include references so the public can then follow up on those things."

And I asked, "So, I think for me, one of the most confusing things is, what level of detail do we include?" I continued with respect to our current task, "We have a lot of information, and we could say a lot more that actually is informative. But we really need some help on how to do that and what is excluded. Are we the ones to exclude information that might be informative? And might we look at assessments from other agencies that include chemical words and data? Because to my knowledge, I was involved with critiquing assessments before, and there was data there. It needs to come probably from a higher level that's associated with written guidance on how we write these assessments in CTP."

I asked, "If it's possible, if we could have a larger discussion about what level of detail is required in CTP's assessments, that would really help."

His response nicely framed our quandary, "Yeah, I agree. And yeah, there's a few points there, that we don't exactly have, we don't have guidance. We have previous examples as precedent, and I think that's part of the challenge with this assignment, is *we* should decide what is appropriate."

A colleague asked, "I think it's still open if we want to add anything or delete some of the information? I don't think we're, you know, wanting to delete references. But I think we would still be open to adding more information."

Supervisor #4 agreed, "Yeah, for sure."

He then ended the call saying, "Beyoncé had made one of the best 'Jolene' covers that I've ever heard. She had Dolly Parton introduce it, and there's still criticism of it that just doesn't understand. So, yeah, I mean, I think that patience is good in this situation, and hopefully it does result in a better product that's more clear. So, with that, so next week we'll touch base on this."

April 9th, 2024

Mr. Hyde then stepped in for Dr. Jekyll. Just three workdays later, at 4:46 pm, Supervisor #4 wrote about our assignment, "The instructions are to look for updated research and only update the text to correspond to the new reference." His email was addressed to all of Team 1, our Division Associate Director, and our Division Director. In that email, he misrepresented our efforts and how he and our Division Associate Director had directed us. He wrote, "There have been discussions about things like adding detailed cigarette-specific secondhand smoke composition data to the Cigar assessment, or abandoning the TTB data that we currently use in lieu of an alternative tobacco product monitoring dataset without evaluating the strengths and differences of these datasets."

We'd been clear that we could add specific information about cigars, and *for years* we'd tried to discuss the differences, strengths, and weaknesses of all available datasets. Staff for some time had communicated specific concerns with our use of industry-reported data that always indicated a market decline, and that therefore always supported the agency's findings of no significant impact. Our branch leadership consistently pointed to introducing new products into a declining market as support for findings of no significant impact. It was clear that evaluating the differences, strengths, and weaknesses of industry-reported data the agency used to always reach findings of no significant impact would call into question at least some of those findings.

The data and how we developed market projections used to support the agency's findings were also concerning, and many in the branch communicated specifically how in meetings and in writing. About ten months earlier, I'd been tasked by Branch Chief #4 and Supervisor #2 to lead a series of discussions, for which I received an award, about that exact set of issues. Supervisor #4 hadn't been a supervisor then. He was a primary reviewer on Team 3 and contributed to those discussions. Back then and very recently, he'd voiced the same concerns.

Preparation for that series of discussions involved several weeks of queries to the entire branch that provided substantiated concerns alongside proposed solutions. But unfortunately, again through the course of that assigned work, problems had been revealed with the credibility of the approaches that CTP used to support its decision making. Changes to improve the credibility of our work would again reveal past cherry-picking that facilitated the granting of marketing orders for tobacco products.

Our Division Associate Director and Division Director hadn't attended those meetings, but they'd been looped in. They'd received the list of concerns slated for discussion and were informed that those discussions ended prematurely when our Branch Chief #3 and #5 changed upcoming meeting agendas to presentations marginally related to our regulatory work. Our Office of Science Director #2 had promised better transparency. He'd said we'd use credible approaches and said there would be better communication. Those regularly scheduled, bi-weekly branch meetings were explicitly advertised to address critical regulatory issues, but they'd been dumbed down and then reduced in frequency to once a month.

We'd raised concerns Supervisor #4 already knew about to Supervisor #4, explicitly in the context of this assignment that he oversaw to update references and text, because we referenced industry-reported data. And because we'd already seen how reticent branch leadership was to remove or even discuss industry-reported data, we'd suggested instead that we add an *additional* dataset or datasets. We were clear that we were uncomfortable using only industry-reported data.

Supervisor #4 chose to misrepresent our efforts, implying we were unprofessional and unscientific by falsely indicating to our leadership that we wanted to make changes "without evaluating the strengths and differences." That was ironic—he wanted to *continue* using industry-reported data "without evaluating the strengths and differences" of that information.

On the same day at 5:13 pm, our Division Associate Director wrote, "The context of the assignment which was discussed in a team meeting was to simply search the literature and see if there is more up-to-date relevant information from some specific references in the Air Quality section. That's the scope of it. I had highlighted some references in an example Cigar assessment to start on. The point is to read and analyze new literature that is relevant to the text and as [Supervisor #4] described, discuss in the team to see if there is room to justify the new reference for potential improvements to that section of the assessment. Having discussions about the findings is the right way forward."

He'd started writing something more myopic but ended mentioning potential improvements to that section of the assessment that we'd been directed to consider.

Then, at 5:29 pm, our Division Associate Director wrote again, "The assignment I gave is about the air quality section on secondhand smoke."

He clarified, "This point of this assignment is to see if we can improve upon the quality of best available science for review and to focus on air quality. The long-term goal of this assignment isn't to force potential findings upon rest of staff. I made it clear that we have to see what kind of info is found first and to analyze it and discuss if it is relevant in a collaborative manner. If, there are any new refs or recommended changes to text then as [Supervisor #4] noted, we'll move it to branch managers to look at. If we were to end up developing it further, branch managers would decide how best to rollout the suggestions and findings from Team 1 to the rest of the branch."

It was good we'd not be forced to use industry-reported data or implement other questionable approaches.

April 11th, 2024

Supervisor #4 then spent the entirety of our next discussion defending our use of industry-reported data. He opened with, "I've been thinking about the questions that have come up with the assessment project and how we can discuss those in the broader context of like, if we're adding stuff to the assessments, how do we evaluate that and make that recommendation in the future. And so, one of those things that's come up that I think is really interesting to talk about is the different monitoring datasets that exist."

He continued, "So, TTB is what we've used for a long time, and I think there's good reasons to do that. It's been collected for a long time. It's managed by a government agency. CTP uses it for determining market share for our user fees and so, while I don't know, and I think some of that institutional knowledge has been lost, it makes sense that when we started doing assessments that that was the dataset that was available and used, and so it got incorporated."

He was markedly more articulate that day. He seemed scripted.

He continued, "But there's also Euromonitor and Nielsen that might be measuring different things. They might also tell us things about particular demographics, if they're focused like Nielsen is focused on convenience stores. And so that might be a different demographic than people who are shopping online or somewhere else. But one thing that has come up, like, we haven't really talked about if the TTB data is good, and I saw that [two colleagues] were starting to have that conversation in the assessment document."

Again, we'd had those discussions before. He knew that those notes in the assessment document didn't constitute the start of those conversations.

He continued, "And so, I was just thinking about that, like, how do we evaluate that? And I realized that we actually have done a little bit of work on that and I just want to share this screen. Do y'all see this figure?" He presented a graph that used TTB data about cigarettes plotted against another dataset and said, "So, you can kind of see that both of them have a decreasing trend over time, but that those are really strongly correlated," and "that suggests that the data is pretty good at tracking the long-term pattern."

The colleague who developed the graph he'd referenced raised his digital hand and said, "A couple of things I wanted to point out with that point you raised with regard to an assessment of the TTB [data]. I've run the same analysis with other industry-reported data." He directed Supervisor #4 back to the graph and explained, "There are so many other measures of use prevalence, and there are so many other datasets available for cigarette usage. But the key here is that I'm not sure if we have looked at anything that is *not* industry reported."

The colleague who developed the graph asked, "You see what I mean?"

Supervisor #4 awkwardly replied, "Yeah."

The colleague responsible for developing the graph then continued, "All I'm trying to get at is that your point is valid to the extent that if we swap out different *industry-reported* datasets, that trend holds. However, we have not looked at it beyond the point of what is reported by industry, and we have discussed quite a bit of this across different points in time. Thank you."

Supervisor #4 flatly responded, "Yeah, that's a fair point."

The number of cigarettes smoked certainly could have been decreasing. Of course that's possible and even likely especially where all sorts of other products were coming on the market. That, however, wouldn't preclude concerns about industry-reported data, especially when other data reported increased *cigar* sales relevant to the agency's determinations and our current assignment.

Our Supervisor #4 had just criticized us to our higher-ups for considering *cigarette data* in the context of what he said should have only considered cigars. He then defended our use of industry-reported data, in the context of that assignment about cigars, using industry-reported cigarette data tested against industry-reported cigarette data.

April 18th, 2024

Supervisor #4 then started our next meeting with a bit of a diatribe, "We can't—we're not going to add all of the information that we come across into this. And so kind of the boundaries of making this something that we can accomplish in the brief amount of time that was dedicated to this is compartmentalizing that amount of information, right? So, there has to be judgment calls on, like, is this really adding a lot of information to the assessment or is this updating the information that's there? The goal for this assignment is to update the information that's there."

That's not how they'd directed us.

He said, "So, the original goal was to look at the citations and look for updated information about the information that's in the assessment. And so, you know, when you find updated information, it might be necessary to change the text as appropriate to match the new reference. So, that's kind of, that was the original goal."

He continued, "We've, you know, we've gone above and beyond. And so, there's a lot of good information. So, when I was looking through this, I was trying to take the information that people had in a lot of cases, I kind of was trying to keep the references and information that was added but also stay true to the information that was in the original assessment."

What?

"Stay true to the information that was in the original assessment"? Now he said we needed to find newer references that matched past claims. One could argue that approach—in this context when we knew more—constituted fraud.

He continued, "So, it's trying to find that balance between like, not adding a ton of new information, but keeping the relevant information. And then when I went through this time, I was trying to reorganize stuff and just kind of like work on the organization of the information and the flow. And so, that, like the considerations there are, like, what is the information that people found and how do we keep that and also be concise and, you know, and that NEPA consideration is discussing things in proportion to their potential impact, that kind of stuff."

He continued, "So, I, you know, I was trying to stay true to the original assessment without adding, with, you know, with the goal being, updating, but not like major renovations. That's not necessarily what happened. And then just thinking, about like overall organization. So, that was kind of my mindset going into this, and also considering that, you know, all of these are proposed revisions, it, you know, depending on, at the next level it would be very easy for them to say, like, 'This isn't really what we had in mind.' So, you know, trying to keep the information within the boundaries of the original assessment is key to that."

Two weeks later, he was replaced by Supervisor #5. He'd said our branch leadership would circle back, and we asked, but they wouldn't engage again about the assignment or the information. The agency has that information and we should see it in publicly posted assessments.

22 | "This should be the preparer's, you know, statement."

I'D ASKED OUR DIVISION ASSOCIATE DIRECTOR during that April 24th, 2024, meeting if primary reviewers who prepared CTP's assessments should agree with the arguments in those assessments. He determined findings for CTP after he reviewed the assessments that I and other primary reviewers in our branch prepared. We'd had that meeting after our CTP Director #2 directed me back to our Office of Science Director #2, who directed me back to him. Leadership that reported to him and that supervised us would not say if we needed to agree. They'd talk around or ignore the question or said they'd check but never followed up. Our Division Director ignored then talked around that question, then directed us to Supervisor #4 who said he'd asked but that he couldn't get a response.

In response to that question, our Branch Chief #3 and #5 had previously said that many years earlier she was part of the small group of FDA employees who developed our assessments. She said that group included four individuals from the FDA's Center for Food Science and Nutrition (CFSAN). CFSAN's approach, associated with products with public health *benefits*, had been applied to tobacco products.

FDA and CTP frequently cited that difference as one reason CTP was faced with a unique set of challenges, and staff frequently indicated an early misstep was to apply approaches developed for products that have public health benefits to products that don't. CTP feigned what my Supervisor #2 described as our "weight-of-evidence approach," purportedly reaching well-supported findings after identifying and evaluating relevant information from multiple sources. Some staff considered CFSAN's approaches questionable, but CFSAN could at least weigh any potential negative effects against public health benefits.

Tobacco products could further benefit from sharing an umbrella with beneficial products. Any FDA-wide request for an exemption, for example from limitations to harmful materials in the products it regulated like medical devices or pharmaceuticals, might default to tobacco products. There might be good reason to allow harmful materials in beneficial products when those products save lives. For instance, exposure to a harmful material that might increase the odds of illness later in life might be tolerated if that individual would soon die without that harmful exposure. One obvious example is chemotherapy, that comes with side effects we choose to risk to survive. Imagine allowing those sorts of exemptions in products with no benefit and that already cause harm.

Anyway, she said, "We came up with how the assessment is written. So, I hope you know that is not one person's decision and it was a collective and it's based on NEPA and NEPA does not have much teeth because the FD&C Act has more teeth, really, for us for our actions. But we still have to do the NEPA and assess the environmental effects."

What?

We clearly weren't doing that, and the question was, "Does the preparer of a CTP assessment need to agree with the arguments in that assessment, or are they just the preparer?" And if you "came up with how the assessment is written" years ago, and it needed to remain static, why hire all of us and tell us to do otherwise? They'd decided years ago what was included and excluded and expected us to copy and paste that to support findings—in every instance—that facilitated marketing orders. However, their names weren't listed publicly as the preparers, ours were. And there was new science to consider, and they kept telling us to consider it. Then they'd backpedal.

"This gets at an important point," I replied. "The problem is, if it's [already] developed and you have new people come in that have additional expertise, new expertise, it won't stay the same. If it is static, that indicates a concern."

It sounded cheesy, but I tried to make an important point while throwing management a bone of understanding and said, "The way laws are implemented changes, that's a wonderful part of our government. That's a wonderful part of NEPA and every law. The law goes on the books and then we decide how to implement it, and clearly that is a struggle for the Tobacco Control Act because everything is happening really fast, right? So, this center is under a lot of challenges."

"One thing that management needs to realize," I continued, "is that we're putting our names on these documents. I have a career. I have credentials. I know the documents that I am putting my name and my credentials on are questionable, and if that's going to continue, maybe one option is to take our names and our credentials off."

But that wouldn't solve the problem.

So, I continued, "Our credentials and our names are put on a public document" and based on "my understanding of our process now, based on my understanding, my experience, and my credentials, I would not agree with the recommendation for a FONSI. I think the agency also has to take that into account when you're using our scientific expertise in these documents to indicate these have been prepared appropriately or, so the agency reflects our expertise in a document we put our name on." I said, "They are thin arguments." I said, "I know professionally that they're a problem. And if I am told that they've been vetted or somebody else agreed in the past, I'll certainly take that into account, but I am also certain that there are concerns about the things we've done."

During that April 24th, 2024, meeting I'd asked, "In the assessment it says, for example, 'The agency considered this information and finds that there are no impacts.' Does the person that prepares the assessment need to agree with those statements? Any statements in the assessment? It doesn't really matter which one. Whatever arguments are presented in the assessment, does the preparer need—is it a requirement that the preparer *agrees* with those? Or are we just *preparing* it?"

Our Division Associate Director said, "If the preparer, they're the preparer so they must have written that. So, if they don't agree with it—then no—you don't have to, this should be the preparer's, you know, statement."

He wouldn't discuss the previously used deceptive approach, but if the current thinking was my thinking, at least I now knew I could reject pressure to use shitty arguments in the assessments I prepared. Because Supervisor #4 and our Branch Chief #3 and #5 were also at that meeting, our branch leadership also knew.

I tried to move forward with that clarity.

23 | Swinging the Lead

CONTRADICTORY COMMENTS FROM LEADERSHIP increasingly necessitated clear guidance about our considerations of air emissions and lead exposure. On April 2nd, 2024, I'd raised concerns about harmful effects to children and other vulnerable groups via secondhand and thirdhand emissions in a memo to our Division Associate Director. He thanked me and indicated that my input was important, and on April 24th, 2024, said he was working on a response. I'd copied others looped in on previous discussions: our Division Director, our Office of Science Director #2, and the CTP Ombudsman.

In that memo, concerns about lead exposure were framed in the context of how the agency justified our findings. For example, the agency dismissed potential harms using the rationale that the same number or fewer units of a new product would replace a comparator product that would be removed from the market. I cited evidence indicating that and other arguments weren't supported.

Even if the arguments used were credible, it was unclear why we hadn't informed the public about important harms and uncertainties. There was no requirement that the agency keep that information from the public. With respect to our obligation to inform the public, the quality of the data and arguments we used introduced even more uncertainty that could instead be countered with transparency. Instead of informing the public, so they could judge for themselves, we thought for them via cloistered and uninformed decision making.

Our Division Director falsely represented what our process constituted all the way to the Office of the FDA Commissioner and ignored and talked around issues directed to him, frequently saying our division of over 100 regulatory scientists answering to more than one statute could "do only one thing at a time." He finally responded to a query about how, during agency decision making, lead was considered. On March 27th, 2024, he wrote, instead, that he and our Branch

Chief #3 and #5 *believed* that products denied marketing orders was "what is going to make a difference."

What?

Although doing good things is no doubt good, that wasn't what he'd been asked. I wasn't there to "make a difference." I was there to do my job, and the query focused on our decision making and regulatory compliance. Understanding how the agency considered lead and air emissions in respect to its determinations to grant marketing orders in the context of the Tobacco Control Act would help us frame what should be reported in assessments prepared for NEPA compliance. Our public health mission required we objectively and comprehensively contextualize agency decision making with what was known and not known. In our assessments, we were to share that with the public alongside the rationale the agency used to grant marketing orders.

This was disturbing. We didn't do assessments for products that were *denied* marketing orders. So, it would be impossible to meet compliance using an assessment that didn't exist. His rationale indicated because we didn't grant some marketing orders, we could ignore lead and other harmful substances associated with products that we granted marketing orders for.

Our assessments informed the public about decision making associated with products that *received* marketing orders, those products the public has access to, uses, and is exposed to as a result of the marketing orders that our agency granted. Those assessments excluded information that was readily available about harms from lead and other contaminants. We also excluded harms associated with uncertainties and information gaps, many we'd already identified. For example, the 2018, 117-page white paper that summarized information about emissions concluded that lacking information across all tobacco product categories precluded informed regulatory decision making. Six scientific disciplines warned about known harmful effects from secondhand and thirdhand contaminant exposures from tobacco products. Risks to the unborn, infants, and youth were communicated, and since 2018, we knew more.

Just the day before, on March 26th, 2024, our Branch Chief #3 and #5—astonishingly—falsely claimed, "We are not actually looking into the human health in our NEPA documents."

This was something I'd not heard before.

She knew about our assignment to update the section of our assessment that, even before our updates, considered human health effects from air emissions associated with the use of tobacco products.

Our Division Director had described CTP's assessments many times, for example, during an October 29th, 2019, meeting, "Deemed Tobacco Product Applications—A Public Meeting." He said, "Importantly, all actions under the FD—the Food, Drug and Cosmetic Act, are also governed by the National Environmental Policy Act, or NEPA, which requires that all actions have an associated environmental assessment or EA or categorical exclusion."

He said, "The need for an environmental assessment or qualified claim of categorical exclusion for each application is not tied to the APPH [Appropriate for the Protection of the Public Health] determination. Instead, an assessment is necessary pursuant to the National Environmental Policy Act or NEPA under 21 *CFR* which states that all applications or petitions requesting agency action require the submission of an assessment or a claim of categorical exclusion. NEPA requires the preparation of an assessment for FDA to proceed with the marketing order for a new product."

He continued, "So, what is an environmental assessment? An assessment is a stand-alone document for the public to understand the government's environmental considerations."

He said, "It's recommended that an assessment include discussion on the following topics," with "air quality" first on his list.

He said, "Potential impacts of tobacco products may include impacts resulting from secondhand and thirdhand exposure from use."

He said, "It's very important to compare the health risk of a new product to comparative products that are likely to be consumed by users of the new product."

He said, "FDA's goal in product regulation is to reduce the public health risk and individual health risk to the user posed by tobacco products available on the U.S. market."

Our Division Director also wrote on March 27th, 2024, that we didn't need to include in our assessments harms associated with a new tobacco product receiving a marketing order if those harms didn't differ from harms for the collective of products making up that class of products.

That was something I'd not heard before.

He was indicating, for example, that we didn't need to consider or share in an assessment, meant to inform the public, outcomes like brain damage in children exposed to lead emissions from cigarettes or ENDS products because other cigarettes and ENDS products emit lead.

If we chose to exclude information that didn't differ within a class of products, we also excluded considering and communicating new information, new considerations of existing information and uncertainty, and newly learned data gaps associated with that class of products. We constantly see the need for new types of considerations and new information about how tobacco products are harmful. However, because a new product receiving a marketing order was part of an existing class of products, we weren't to consider or include any mention of any of that in our assessments?

"I do not think that we have encountered such a situation yet," he continued, "some kind of specific harm to humans that is different from products of the same class." But how would we know that if we didn't consider thresholds, if we weren't comparing exposure levels from those products alongside levels where harm might occur? Our approach precluded any such encounter.

Each layer of the onion revealed more stink.

On April 16th, 2024, he provided a different explanation and pointed to a resource that seemed to shed more light on the situation. He wrote, "In general, SE Toxicology review is driven by ingredient changes [in the new product] as compared to the predicate [comparator] product [that is already sold]."

I was following. "SE" refers to products considered via the substantial equivalence pathway where a marketing order is granted if the new product is similar to a product already on the market and no new public health concerns arise from marketing that new product. His mention of the "Toxicology review" referred to what the other branches in our division did. In my branch, we did the Environmental Science review, and some of that internal data we'd hoped to use in our assessments was used in and generated via the Toxicology review.

He continued, "If ingredient changes could potentially cause a new product to have higher HPHCs [Harmfully and Potentially Harmful Constituents] than the predicate product, CTP asks for relevant HPHC data. However, in practice, there is a list of 20 HPHCs for which there are validated analytical methods," and he provided the link to that list of constituents.

The link indicated that from 93 HPHCs, a selection of 20 were considered for some products:

- For combustible cigarettes, the agency considered 18 of those constituents: acetaldehyde, acrolein, acrylonitrile, 4-aminobiphenyl, 1-aminonaphthalene, 2-aminonaphthalene, ammonia, benzene, benzo[a]pyrene, 1,3-butadiene, carbon monoxide, crotonaldehyde, formaldehyde, isoprene, nicotine, NNK, NNN, and toluene.

- In smokeless tobacco products, the agency considered nine: acetaldehyde, arsenic, benzo[a]pyrene, cadmium, crotonaldehyde, formaldehyde, nicotine, NNK, and NNN.
- And in roll-your-own tobacco and cigarette filler products, the agency considered six: ammonia, arsenic, cadmium, nicotine, NNK, and NNN.

He then wrote, "FDA is currently exercising enforcement discretion to require reporting about these 20 constituents."

Enforcement discretion refers to FDA's decision to not fully enforce the law.

The FDA is required to publish the quantities of 93 HPHCs in tobacco products "in each brand and subbrand of tobacco product, in a way that people find understandable and not misleading."

That hasn't happened, but would no doubt help inform the public as well as CTP scientists. Regardless, the agency should consider those harms and potential harms as part of its determinations to allow products to be sold in the U.S.

Especially when it doesn't, that is the very sort of disclosure we should've included in our assessments. We might've said something like, "During the FDA's evaluation of [product name], the FDA considered nine of 93 constituents found in some tobacco products or produced during the use of some tobacco products that the FDA has determined are harmful or potentially harmful." If more or fewer HPHCs were considered, that could be clarified. Constituents inapplicable to the product, for example, those emitted by only combustible products therefore inapplicable to non-combustible products, could be listed as such. Information already compiled in usable formats could be provided about the harms of relevant constituents that were considered and that weren't. How those knowns and unknowns and any other considerations informed the agency's decision making could be included.

Then later that day, our Division Director wrote, "Unlike any other part of FDA, the main health risk evaluation conducted by CTP is a risk assessment relative to other tobacco products on the market—not a safety assessment or a risk assessment that compares the risk of a drug to acceptable risk determined by the risk of a disease (such as in the case of cancer drugs). Pathways to tobacco product authorization are mostly based on comparisons to other tobacco products on the market, although this may change sometime in the future if we are able to finalize product standards. So, currently, we may authorize products that contain levels of lead that would be unacceptable in food or drugs."

That's something else we might've disclosed to the public. We might've said, "The FDA may authorize tobacco products that contain levels of lead and other harmful constituents that would be unacceptable in food or drugs."

And we might've needed to preface that with the broader context. We might've said something like, "The FDA is required to publish, but has not yet published, quantities of 93 harmful or potentially harmful constituents found in tobacco products, in each brand and subbrand, in a way that people find understandable and not misleading. The FDA's decision to allow a tobacco product to be sold in the U.S. is not a safety assessment or a risk assessment, for example, that considers the risk of disease or side effects. The evaluation conducted is of some risks relative to other tobacco products on the market, for example, where new products can result in the same harms associated with products already sold in the U.S. However, the FDA is currently exercising enforcement discretion and considers a selection of those 93 constituents."

And because we were to use plain language that is easily understandable to the public, we might've said something like, "Enforcement discretion means the agency has chosen to not fully enforce the law," alongside some rationale for that decision. My guess had been that was because of difficulties amassing and considering the science associated with what is known and not known about each of those 93 constituents. That said, we know enough about some constituents excluded from the list of 20 that were sometimes considered.

Lead is on that list of 93 constituents that the FDA determined are of concern in tobacco products. It is listed there as a carcinogen, as a cardiovascular toxicant, and as a reproductive or developmental toxicant. However, lead and many other harmful constituents were excluded from the list of 20 constituents selected as part of the FDA's enforcement discretion, as part of the FDA's decision to not fully enforce the law.

So, we hadn't "encountered such a situation yet," "some kind of specific harm to humans that is different from products of the same class," because in many cases we weren't looking. If we didn't bother looking at the amount of lead in the new product and in the comparator product, how would we know if there was more lead in the new product compared with the comparator product? And what about those new products for which there was no comparator product to compare it with, with respect to those 20 HPHCs or all 93 HPHCs?

And we could look. Claims that analytical methods preclude credible considerations of lead and other constituents are false.

Our Office of Science Director #2 joined CTP in 2023. Like me, he hadn't been at CTP when the 117-page white paper had been completed, so I shared the paper with him. On April 16th, 2024, he wrote, "I think the key for emerging issues is to see how they play out within our existing authorities. As we know, if it's not within our authority, it can take a long time to change. While there is certainly evidence from previous studies of lead in cigarette smoke, it's a different type of risk exposure than kids eating tainted applesauce that has acute effects in short order on kids' health. Of course, we also must aim our finite resources where we can have the biggest impact on public health. That means we can't tackle everything at once or with the same level of effort."

What?

Who said we had to tackle everything at once? My aim had always been to prioritize, otherwise I'd be talking about a lot more than air emissions and lead. And harmful effects of lead have been recognized for millennia.

I wanted to tread lightly. I didn't want to insult him, but I needed to let him know that lead was far from an emerging issue. Did he really not know that? He mentioned "evidence from previous studies of lead in cigarette smoke," so it seemed like he knew. Why would he say otherwise? And what would be worse, for our Office of Science Director #2 to think lead was an emerging issue or to pretend it was, especially in this context. Was this ignorance or gas lighting?

In my response, I simply clarified that "lead is not a new issue" and repeated that the purpose of our assessments was to inform the public.

He certainly knew. He described his experience during that October 24th, 2023, public meeting, "Premarket Applications: Opportunities for Stakeholder Engagement." He said, "My dissertation was actually looking at the role of workplace smoking policies and cigarette excise taxes on adult smoking behavior, and that was in the mid-90s, so before it was cool to be in tobacco research. So—and ever since then, my research has focused on just that: looking at smoking behaviors, later vaping behaviors, among adolescents, young adults, and adults."

He said, "I consider myself a public health researcher."

He said, "I worked with many health departments across the country, worked with the Office on Smoking and Health starting in the mid-90s, along with the National Cancer Institute. And then, soon after FDA's Center for Tobacco Products was formed in 2009, I started working with them in 2010 when the center had a size of about 25 people, and I worked with them continuously through 2023 in March when I transitioned to this role at CTP."

Now he knew better, but then he'd said, "Despite working with the Office of Science and others at CTP for so many years, I really didn't know a lot about sort of the inner workings of the Office of Science, and I was really impressed when I joined CTP to learn about the complex and rigorous application and review process."

He said, "And I think at the core of what I bring to the Office of Science is the idea of scientific evidence driving decision making."

Great idea!

On April 22nd, 2024, our Branch Chief #3 and #5 again talked about how we addressed air emissions, across all the products that CTP regulated, in our assessments. Now she said, "Any other information that we can add to the air emission [section] from users, from use of these products, is going to be difficult to address because we don't have actual robust data and information on that one. Everything that's, there is, we're using literature cited on those, and we rely on that for the use. We're focusing more on the impacts from the disposal, because really from use, most of the impacts are from disposal after use of these products. And actually, mostly it's from littering."

What? Over 40,000 Americans die each year—die—every year—just from secondhand emissions from tobacco product use.

I know it's not impossible, but I know of no one who died from an incident involving tobacco product litter, or from any other form of tobacco product or packaging disposal, with one exception—fires. When I was a youth, I threw a burning cigarette out my car window. It blew back in and caught the backseat on fire. Thank goodness for fire retardant materials, it quickly went out. I didn't die but others have in fires and wildfires caused by tobacco products.

And of course, tobacco product waste is a source of hazardous contaminants. I'd for years been involved with efforts to understand those contaminants, focusing mostly on microplastics but also a bit on batteries. Our ability to evaluate those exposures paled in comparison with what we could provide about human health and air emissions.

Our division brimmed with toxicologists, who generally in these sorts of rolls look to prioritize the most damaging harms associated with the best data. So, it was surprising that we'd decided to focus instead on disposal—and that was something I'd not heard before—particularly after we'd just shared new means to estimate some specific air emissions associated with product use.

It didn't say anywhere that the agency decided not to seriously consider tobacco product use. We'd need that directive in writing. I couldn't just leave it out based on unsubstantiated word of mouth, and it increasingly seemed like concerns inconvenient to address were talked away via these unwritten directives.

Discussions of product use were historically ubiquitous, even in public communications. For instance, in August of 2016, CTP's Associate Director for Science Policy, who in 2024 was our Director of the Division of Regulatory Project

Management, presented to industry about the need to evaluate "Environmental issues relating to use and disposal from use of the tobacco product." On July 17th, 2024, she again confirmed that we needed to seriously consider product use, as well as data quality issues, public disclosures, etc., via policy excerpts she added to an email she wrote:

21 CFR 25.40(a) which states, *"...The assessment shall focus on relevant environmental issues relating to the use and disposal from use of FDA-regulated articles and shall be a concise, objective, and well-balanced document that allows the public to understand the agency's decision..."*

21 CFR 25.40(e) which states, *"The agency evaluates the information contained in an assessment and any public input to determine whether it is accurate and objective, whether the proposed action may significantly affect the quality of the human environment, and whether an environmental impact statement or a FONSI will be prepared..."*

Anyway, our Division Associate Director hadn't followed up on his April 24th, 2024, promise to address the memo he'd received on April 2nd, 2024. I waited a bit, then on May 29th, 2024, I emailed our Division Director and copied our Office of Science Director #2. There was no response after a week, so I emailed our Deputy Director for Regulatory Management on June 5th, 2024. On the same day, he responded about the 117-page white paper.

He wrote, "Thank you for passing along this white paper. I believe that this is one of the outcomes of a working group focused on understanding the potential effects of secondhand and thirdhand smoke upon population health."

What?

He *believed*?

He wrote, "At the end of the day, because this is a white paper, I believe it was shared with the senior managers of the individual disciplines. It was the responsibility of the discipline leads (Environmental Science Branch Chief in your case) to determine if and to what extent the information provided should affect the discipline reviews. In 2018, we had not yet received any deemed product applications and were focused on SE and EX applications."

He said, "It ultimately is up to the discipline reviewer to be aware of the literature and to include the relevant information in their reviews. As this was all published literature, the lack of inclusion in assessments likely reflects the lack of concern expressed by the Environmental Science Branch reviewers with respect to the summary presented in the white paper. If you are concerned and wish to include this information in the assessments that you are responsible for, you are encouraged to add it and engage with your leadership in the finalization of the review."

Staff knew that he knew not only about that working group and that 117-page white paper, but also about concerns associated with lead exposures across consumer products, including all tobacco product categories. He wasn't ignorant of any of that, and he knew staff knew that. His experience and expertise were frequently chronicled in internal meetings and externally. One of those accounts is captured in that October 24th, 2023, public meeting, "Premarket Applications: Opportunities for Stakeholder Engagement":

[Our Deputy Director for Regulatory Management] *joined CTP as a Chemistry Review Scientist in 2015 and progressed to managerial positions until he served as the Associate Director of the Division of Product Science. In his position, he oversaw the evaluation of the composition and design of tobacco products. In addition, he served on the leadership team where he was involved in chemical, microbiological, and engineering research on tobacco products resulting in numerous publications in peer reviewed journals.*

In 2020, [he] *became the Deputy Director of CTP's Office of Science where the leadership team in the Office share responsibility for reviewing tobacco product applications.* [His] *particular focus is on substantial equivalence (SE) reports and exemptions from substantial equivalence (EX) requests.*

In addition, [he] *participates in aspects of premarket tobacco application (PMTA) product review, provides scientific support for regulations and guidance, evaluates the knowledge base for regulatory decisions, and carries out research to fulfill the gaps in scientific knowledge related to tobacco product regulation.*

Before his tenure at CTP, [he] *worked for over 20 years at the United States Pharmacopeia Convention, USP, including ten years as the Vice President of Standards Development. USP is a non-profit standard-setting organization for the pharmaceutical industry. In his role,* [he] *authored and contributed to hundreds of pharmaceutical standards and chapters with broad industry impact, including a revised standard for metal impurities* [lead is a metal] *in pharmaceuticals. In his role, he traveled extensively and presented at hundreds of national and international symposia and training sessions throughout the world.*

It wasn't great to hear reviewers blamed again when he pointed to what he characterized as our "lack of concern" as the reason information from that 117-page white paper had been excluded from CTP's regulatory science work. We knew of no efforts to distribute or discuss that paper or its contents. Newer staff hadn't seen it, so how could we have been unconcerned about its conclusions? I'd been at CTP since early 2021 and had just recently learned about it from a colleague concerned it had been buried. Indicating that a single reviewer should've reached the same conclusions that required dozens of scientists from six disciplines three years to reach because "this was all published literature" seemed disingenuous. If leadership knew about concerns like those in that paper, they should alert staff in each applicable discipline to consider that in CTP's regulatory science work. He'd said Branch Chief #1 decided, so how had he determined reviewers weren't concerned? This sounded like passing the buck.

Why on earth had the paper been produced in the first place if what was learned would be ignored? What changed since the workgroup had been assembled, sometime before 2018, that meant the impetus for all that no longer mattered? We certainly had no grounds to dismiss concerns about lead, or about air emissions, or about the many other things in that 117-page white paper.

And reviewers had absolutely tried to add information about lead and emissions to our assessments. To say those omissions likely reflected reviewers' lack of concern was laughable considering leadership's efforts to wipe the concerns of reviewers from CTP's regulatory science work. Consider just our effort to update the air quality section of our assessments that had gone dark.

He'd explained, "In 2018, we had not yet received any deemed product applications and were focused on SE and EX applications." However, leadership knew the information, recommendations, and warnings included in that paper applied across SE, EX, and PMTA applications. For example, the focus in that 117-page white paper of just one involved discipline, Chemistry, was to identify Harmfully and Potentially Harmful Constituents (HPHCs) in secondhand emissions and exposures from ENDS and any heated or burned tobacco product. That also included waterpipe tobacco, cigars, and cigarettes. And concerns about mainstream and thirdhand emissions were also discussed. Concerns about lead were included. And he'd been a Chemistry review scientist before he'd become a Technical Project Lead responsible for considering all the science, collectively, from all disciplines. And Chemistry's part of that white paper even cited our Office of Science Director #1 as supportive of the entire effort. So, he and other Office of Science leadership knew about that white paper.

Regardless, that knowledge in this role as our Deputy Director for Regulatory Management might've provided the impetus, then, to ensure that information was considered going forward.

At the end of the day, however, it seemed like we had a plan. Our Office of Science Director #2 chimed in and indicated if "sharing such information wasn't [previously] all that central to the assessment. That doesn't prevent reviewers from doing it moving forward."

Great!

And despite his apparent ignorance about air emissions and lead and that 117-page white paper, our Deputy Director for Regulatory Management had also written, "If you are concerned and wish to include this information in the assessments that you are responsible for, you are encouraged to add it and engage with your leadership in the finalization of the review."

This aligned with what our Division Associate Director had said on April 24th, 2024, about statements in our assessments being our own. So, it was good to see that new consistency with respect to our roles.

Also on June 5th, 2024, our Office of Science Director #2 wrote that, "We could develop a summary statement of children's exposure to secondhand emissions that could be repeated in assessments and updated as that information changes at the population-level. That could be based on an updated literature review that reviewers rely on. To me that could meet the obligation. Of course, from my perspective, that would do little to educate the public as very few would be looking at our reports for this information when EPA and CDC have more accessible information on this topic."

The point about meeting our regulatory compliance wasn't getting through. Although we were to consider cumulative impacts as well, our compliance required we parse each analysis by each federal action. So, we needed to quantify exposures associated with each product, not only address changes at the population level. And we were the agency that regulated tobacco products, so the obligation to inform the public came with each regulatory decision that *we* made.

We'd used publicly available information from other agencies before, with respect to things like manufacturing waste and packaging materials. But we needed to add information about impacts associated with the use of tobacco products. And although it was normal for one federal agency to consult another with respect to their NEPA compliance, unfortunately, our requests to do that were always ignored or denied. I know of no instance where the section of our assessments where we were to provide "A Listing of Agencies and Persons Consulted" said anything other than "none" or "not applicable."

Regardless, this seemed like progress!

Unfortunately, I never saw efforts to develop that summary statement or anything like it. The agency has this information and we should see it in publicly posted assessments.

24 | "Complexities around regulatory logistics"

AND THERE WAS MORE THAT WE EXCLUDED. For instance, four scientists for over a year described impacts to critical wildlife habitat burned when tobacco products are littered. NEPA required we consider other statutes like the Endangered Species Act, but we only *mentioned* wildfires in our assessments. Therefore, that effort focused on habitats where endangered birds, bugs, plants, and so on, live, and would better inform the agency's decision making and the public about risks to protected species. The analysis used recent, credible data and focused on disposal, something we'd increasingly been told to consider.

That effort had been supported by all leadership since its inception. Team 1 Supervisors #2, #3, #4, #5, and #6, Team 3 Supervisors #1, #2, and #3, Branch Chiefs #2, #3, #4, #5, and #6, our Division Associate Director, and others supported the effort and commented on the draft. They received updates about that work and saw weekly meetings for many months on our calendars. We'd satisfied several rounds of internal reviews. Little remained before a manuscript would be ready to submit for publication and our assessments would be better informed.

Instead, in late July of 2024, Branch Chief #7 ended the project with only a message in a comment bubble, "Thank you, team for the time spent working on this proposal. While it is interesting, it does not fall within the branch's research priorities at this time. I am unable to move it forward. Thank you all for your hard work."

Regardless of whether the manuscript would be published, the agency has this information and we should see it in publicly posted assessments.

Early on, I'd assumed we didn't consider some things in our assessments because the science wasn't there yet, but that wasn't the overriding rationale.

Leadership characterized our relationship with the tobacco industry as contentious. We knew which of our practices gained their ire, because those were things they'd sued us about, and there were a lot of lawsuits. We knew which practices they'd tolerated, because when we did those things they left us alone. Therefore, to prevent litigation that was expensive and time consuming for the agency, with outcomes that were often expensive and time consuming, we could choose to do the sorts of things they'd tolerate.

Using the most up-to-date and best information in robust reviews of the evidence, by comparison with what had come before, would reveal past blunders. More informed considerations would also reveal concerns alongside uncertainties and data gaps that would require we develop environmental impact statements. That would've required public comment, and we'd need to respond to those comments. That would all slow marketing orders and also increase the likelihood that our mounting credibility issues would be discovered.

So, maintaining the status quo was safer because updating the science, improving our approaches, or making corrections would chum the interrogatory waters. Instead, we filled our assessments with just enough not-too-bad effects to feign compliance.

That said, we excluded other things that likely wouldn't have caused delays or drawn litigation from the industry. For instance, we should see considerations of greenhouse gas emissions in assessments produced during the Biden administration. That was something we were required to do. CTP had complied and quickly removed those considerations when the first Trump administration rolled back that reporting. However, CTP didn't comply with the Biden administration's January 20th, 2021, Executive Order 13990 to bring them back.

Our Branch Chief #1 tasked staff to do that, but efforts stalled. After she left in early 2023, leadership wouldn't commit when asked if we could implement what we'd developed or revert to what had been done before. Instead, for years under the Biden administration, we continued what the first Trump administration had wanted. But when the time came, our Branch Chief #7 provided immediate directives to abide by the second Trump Administration's Executive Orders.

Why didn't we follow Biden's Executive Order? CTP's defense against any possible litigation when we made those changes could be that we were abiding by an Executive Order.

Some at CTP were sympathetic with the industry or with one administration or the other. Some were angry at Trump and some just didn't want to do what Biden said we should do. Additionally, a good percentage of Americans are addicted to tobacco products, and CTP didn't consider that conflict of interest among its staff. Those biases were apparent, as well.

So, what we did wasn't only influenced by regulatory capture, where an agency overseeing an industry is influenced by that industry, for example, owing to fears of litigation. Our work was also influenced by personal preferences and politics. There'd certainly be individuals on one side or the other at all levels at CTP and changing tippy-top management along with each new administration might not change the behavior of mid-level management directing work on the ground.

However, signs of regulatory capture were apparent everywhere. One example involved cigarette filters. People believe filters filter things in ways that make smoking with filters less harmful. But there's evidence they make things worse, for example, when users inhale harmful constituents more deeply. Filters are the most discarded item on the planet, and a persistent source of hazardous waste that additionally degrades into microplastic pollution. Furthermore, users also inhale any microplastics released when they use filtered products.

CTP leadership said filter design changes or product standards were outside CTP's regulatory authority. However, staff looked into it and learned they weren't. Those efforts were sidelined because CTP thought the industry would push back. Efforts the industry was more likely to tolerate or partner on were prioritized.

Other failures to apply the agency's regulatory authority prevented the agency from protecting the public from or informing the public about harms associated with lead exposures and emissions from tobacco products.

On March 8th, 2024, a colleague had written to our CTP Director #2 about lead exposures from tobacco products. He wrote, "I write this with some trepidation, given your vast expertise on the topic. I write hoping to focus your attention on environmental tobacco smoke (ETS) as a source of lead exposure for children, and to spur some action for ways to reduce such exposures which disproportionately affects minority and low income children. I am not sure if this has received much attention within CTP (see attached Office of Science white paper from 2018 on environmental tobacco exposure), and [CTP's Senior Science Advisor] is not aware of efforts related to this either."

The white paper he attached is the same 117-page white paper previously mentioned.

He continued, "A recent review article reported that sidestream smoke from each cigarette emits 17μg of lead. Currently, CDC uses a blood lead reference value of 3.5 micrograms per deciliter (μg/dl) to identify children with high blood lead levels. As far as I'm aware, the 2006 Surgeon General's report on the topic does not address this and at present ETS/secondhand smoke is not even listed as a source of lead exposure by CDC! A quick glance of research studies on the topic (examples spanning 2003-2023; here, here, here, here, here, and here) and a systematic review identify the association between ETS and elevated blood/saliva lead levels, with inhalation as a pathway of exposure for children and disproportionate lead exposure for black children."

He provided links to the recent review article, the 2006 Surgeon's General's report, the six example 2003-2023 research studies, the systematic review, and referenced CDC information about lead levels and sources of lead.

He continued, "What prompted me to write directly to you was FDA's alert yesterday on elevated levels of lead in cinnamon. I also wanted to bring it to your attention as I sense a huge opportunity to advance health equity and environmental justice through targeted educational programs and potential research projects for funding."

Our CTP Director #2 responded the next day, "Appreciate you reaching out on this. As you know, secondhand smoke related research is very near and dear to me—my dissertation focused on behavioral surveys and environmental assessments of exposure in multiunit housing facilities. There's still a lot more work to do on this front nationally, particularly with regard to comprehensive smoke free policy adoption in public indoor settings in the many states that lack these commonsense laws—as well as ongoing work in non-public settings (e.g., voluntary smoke-free home rules and smoke free multiunit housing policies), particularly to address pediatric exposure."

He continued, "The science is fairly robust in this space, so my druthers is that a comms approach would likely be most impactful. Although the primary rate limiting factor around work in this space is really our ability to relay the public health 'so what'—i.e., the single overriding health communication objective."

"When it comes to secondhand and thirdhand exposure," he continued, "the established recommendation is smoke-free policy implementation, which is unfortunately not something that FDA has authority to do. Our authorities cover constituents, but given the myriad and well documented harmful ingredients in tobacco products—and the complexities around regulatory logistics—this wouldn't be a viable pathway for success. Fortunately, states and localities do have authority to implement smoke-free policies, and various federal partners engage actively with them in this space, inducing CDC and the National Cancer Institute."

He ended with, "That said, it might be worth engaging with CDC, the National Cancer Institute, and perhaps even EPA—to ascertain whether there's appetite for some opportunity for comms in this space. At the least, updating web content or otherwise showcasing lead in a way that grabs the public's attention could be something FDA could partner with these agencies on to best reach the target audiences with the best recommendation to effect meaningful public health change. Toward that end, I'll plan to raise with CTP's Office of Health Communications and Education once we've got a permanent director in place, and perhaps we can look at options with coordinated Office of Science engagement. We recently did a nice collaboration with EPA on e-cigarette disposal, which resulting in a nice public fact sheet."

My queries were a little different. I focused on our compliance with respect to informing our decision makers and the public about lead exposures associated with tobacco product use. In that context, our CTP Director #2 had written that concerns about information excluded from our assessments should be broached within the Office of Science.

In that context, it bears repeating what our Office of Science Director #2 wrote, "I think the key for emerging issues is to see how they play out within our existing authorities. As we know, if it's not within our authority, it can take a long time to change. While there is certainly evidence from previous studies of lead in cigarette smoke, it's a different type of risk exposure than kids eating tainted applesauce that has acute effects in short order on kids' health. Of course, we also must aim our finite resources where we can have the biggest impact on public health. That means we can't tackle everything at once or with the same level of effort."

I'd seen no evidence that lead exposures from applesauce had been compared with lead exposures from tobacco products. But I don't know why that comparison was important. If the effects of lead in applesauce were worse, that wouldn't preclude FDA's consideration of lead exposure from tobacco products. The only change I was trying to make was to better inform the public in our assessments meant to inform the public about tobacco products.

On the same day, our Division Director wrote, "Pathways to tobacco product authorization are mostly based on comparisons to other tobacco products on the market, although this may change sometime in the future if we are able to finalize product standards."

The FDA defined a tobacco product standard as, "A rule that sets requirements for how a tobacco product can be made. It is based on rigorous evaluation of scientific evidence and FDA's determination that the product standard is appropriate for the protection of the public health. For example, a product standard could include reducing or eliminating certain ingredients or chemicals in tobacco products or limiting nicotine levels of tobacco products."

The agency said product standards have "the full force of the law and must be adhered to," and further describes a product standard as "a powerful regulatory tool because they have the ability to significantly reduce tobacco-related disease and death by addressing the addictiveness, toxicity, and appeal across a whole product category."

On June 5th, 2024, our Office of Science Director #2 wrote, "I do want to note that I do not believe that exposure to secondhand smoke (SHS) among children would change if a product is deemed substantially equivalent (SE). That is, if the new product is SE, then that would imply that SHS exposure would remain largely the same. While the baseline level of exposure could be of concern (I topic I'm familiar with as I spent many years studying SHS exposure and SHS policies for CDC, States and EPA), I don't follow how it's within our regulatory authority to effect changes to SHS exposure."

In response, I clarified, "This information should be included in the assessment, regardless of the agency's decision to grant a marketing order. While it might not be in the agency's regulatory authority to affect changes in SHS exposure, it is the agency's obligation to inform the public about exposures and about the agency's decision making."

The focus on informing the public was again sidetracked, and it *was* within our regulatory authority. We could develop a product standard that reduced or eliminated lead in emissions—therefore reducing or eliminating exposures—by reducing or eliminating lead in tobacco products.

In October of 2024, an audience member at the American Public Health Association Meeting, in Minneapolis, Minnesota, asked our CTP Director #2 about CTP's work with secondhand emissions. His quick response was only that "Secondhand emissions are outside CTP's regulatory authority." He didn't mention "when it comes to secondhand and thirdhand exposure, the established recommendation is smoke-free policy implementation, which is unfortunately not something that FDA has authority to do. Our authorities cover constituents, but given the myriad and well documented harmful ingredients in tobacco products—and the complexities around regulatory logistics—this wouldn't be a viable pathway for success." He might've also clarified that we didn't bother considering most of the constituents we were mandated to disclose.

My colleague and I focused on known harms from lead exposures that the FDA is obliged to address. He focused on efforts to reduce lead exposures and I focused on informing the public about lead exposures. Our CTP Director #2 said fairly robust science was available, but that implementing smoke-free policies was not something the FDA had the authority to do. However, smoke-free policies don't reduce lead exposures where products are or were used.

He did say our authorities cover constituents. Lead is one of those constituents. Via its ability to implement product standards, the FDA might be the best or the only authority with a real mechanism to reduce lead exposures from tobacco products that other industries have pivoted to accommodate.

I never saw a prioritized list of constituents, so I can't comment on where lead is on that list. In January of 2025, the agency proposed a product standard that would limit the amount of nicotine in some combustible tobacco products, but not lead or other harmful constituents. Those exposures might *increase* if people use *more* lower nicotine products to satisfy their addiction.

If the complexities around regulatory logistics preclude a viable pathway for success via the FDA's existing authority to implement product standards for something like lead, wouldn't the agency want to ensure those harms are communicated in its assessments? Would the agency want to at least list those harms while being transparent when the agency didn't actually look? The agency has this information and we should see it in publicly posted assessments.

25 | Ignore Small Text

IN JUNE OF 2024, I received a new assignment to evaluate a group of waterpipe tobacco products also known as *shisha* or *hookah tobacco*. The applicant requested marketing orders via the SE pathway in May of 2020, and we were behind schedule. So, I dug in right away.

During my review, I noticed that each product was described on the package several times as "pipe tobacco" and as "hookah tobacco." On the back of the package, it said, "Ingredients: Pipe Tobacco."

While other primary reviewers, like those doing Chemistry, Microbiology, or Engineering reviews, might've limited their evaluations to certain parts of the information applicants provided relevant to those disciplines, the Environmental Science reviews I did required I look at everything submitted and otherwise available, within reason considering our timelines. That ensured we'd likely considered all potential effects, and it meant we saw more than others. Therefore, our training required we raise any potential issues we came across, even if they fell outside the purview of our discipline.

This time, the issue was relevant to mine *and* other disciplines. Different emissions are produced when the same product is used as waterpipe tobacco compared with burning it in a pipe. Beyond mainstream exposures, there'd also be differences in secondhand and thirdhand emissions. However, these products had so far been evaluated only as waterpipe tobacco. Additionally, other products regulated by the FDA list on the package after "Ingredients:" *all* the ingredients in the product, in order from those making up the most to the least of the product in the package. So, another serious concern raised was that the mislabeled package misled consumers into believing the only ingredient in those products was tobacco. It wasn't. Users hoping to avoid or know about exposures to those other unnamed ingredients were falsely led to believe they weren't in those products.

Several solutions were initially discussed. Branch Chief #7 agreed there was a concern because there would be different emissions and therefore different exposures that hadn't been considered or communicated. She believed we should reach out to the applicant. They might've been asked if they could provide a consistent description on the package. We could've also evaluated those products as pipe tobacco.

Instead, my Secondary Reviewer #2 said it wasn't a concern because it hadn't been raised before.

What?

Before when? This was the first round of reviews for these products, and we'd had the assignment for maybe a week.

I disagreed and was then told the issue wasn't a concern because another primary reviewer from Team 2 in my branch, who was reviewing a set of similar products with the same labeling, didn't raise the issue. However, he didn't *notice* the issue. My Secondary Reviewer #2 said, "We just happened to catch it when no one else did."

Obviously, that reviewer wouldn't have raised an issue he wasn't aware of.

I disagreed and was then told it wasn't a concern because the other primary reviewer didn't *think* it was a concern. His rationale wasn't provided. When I asked, my Secondary Reviewer #2 said, "He just didn't think so."

I indicated that our Branch Chief #7 had already confirmed concerns about different mainstream, secondhand, and thirdhand emissions.

Our Deputy Director for Regulatory Management then chimed in. He concluded it wasn't a concern because the moisture content would make it hard to burn those products in a pipe. But we knew people could burn waterpipe tobacco in a pipe. If the label said it was "pipe tobacco" and we granted the marketing orders with that labeling, any bad outcomes would—fairly—be on us.

I mentioned that I was a youth once, that I recognized the capacity of determined youth, or to be fair of anyone, to make things burn. Regardless of the moisture content, people did it anyway. I reminded them and our Office of Science Director #2 of evidence that happened and about potential associated health concerns. I reminded them that was one reason there was a concern, because burning some higher moisture products in a pipe, instead of a waterpipe, might be associated with the production of emissions that are more harmful. I even popped into a tobacco shop and bought a brick of waterpipe tobacco and a little corn cob pipe. Indeed, waterpipe tobacco can burn in a pipe.

Our Deputy Director for Regulatory Management then concluded it wasn't a concern because the font size was small.

What?

This was something I'd not heard before.

I hadn't seen guidance indicating that text was disregarded if it was small. If that was the case, we'd probably need to tell the public. But that didn't seem like a legitimate reason. CTP believed, in other contexts, that small text was informative. For example, it had been appropriate to include in small text below a table in a confidential appendix an apparent explanation of fuzzy math "to cover our butts." However, in *this* case, consumers and CTP were to ignore text because it was "small."

However, the text wasn't small! It was 14-point font or larger.

I suspect that—of all agencies—the FDA shouldn't ignore small text, especially when it comes after "Ingredients:" on the product label. The FDA directs the public to much smaller text to understand health information and ingredients on nutrition labels. And as a group of scientists required to scientifically evaluate the outcomes of tobacco product use, this seemed like a normal thing to not ignore.

Primary reviewers uncomfortable doing questionable work were placated with the option to delete from their reviews a statement attesting to its accuracy. Therefore, in a comment bubble in my draft review for these products, on June 17th, 2024, I wrote, "Because I do not believe this review is accurate based on my understanding of our purview and process, I am deleting the statement, 'I attest the contents of this review and evaluation are accurate to the best of my knowledge.' I understand that we CAN do that in Environmental [the Environmental Science Branch], and that some reviewers often do, but I prefer not to and believe we are creating problems with our regulatory compliance."

My Secondary Reviewer #2 responded, "If my understanding is correct, now I guess, we shouldn't make any changes to the template unless instructed from the powers that be so I have rejected that deletion, I understand that you feel uncomfortable about the review processes in its current form. The work is still to the best of your knowledge at the time."

What?

I responded, "It is not."

I asked, "Why have some reviewers been allowed to delete this statement and I am not? How can someone else tell me work is to the best of my knowledge, especially when I've raised the concerns to follow?"

I said, "This seems inappropriate and I am increasingly uncomfortable with the pressure to sign my name to work that is not backed by agency policy."

Our Branch Chief #7 then wrote, "Office of Science policy for reviews includes using this template and the review job aid advises not to make changes.

I would like to be consistent in that regard. However, if you feel the statement doesn't represent your view of the contents in the review and evaluation, you are welcome to remove it."

I waited a few weeks to see if we could do things aboveboard, but that wasn't happening. So, on June 28th, 2024, I let her know, "I will now take you up on your offer to remove the statement indicating on the cover page that I attest to the accuracy of the scientific review."

Crazy! Branch Chief #7 was supportive of our ability to *not attest to* the accuracy of our work.

Then on July 9th, 2024, our Office of Science Director #2 chimed in, "While there may be quantitative differences in emissions between waterpipe and pipe tobacco use, it's hard for me to imagine that the impacts would be so substantial to have meaningful impacts on public health. To me it's analogous to statistical significance vs. practical significance."

He said, "So, rather than focus on the labeling, the review was done based on the applicant characterizing the products as waterpipe. [Our Deputy Director for Regulatory Management] also said that waterpipe tobacco would not work well as a pipe tobacco because of the amount of humecant in waterpipe tobacco."

He concluded, "In the end, we need to classify it as one or the other. It seems like the choice here, following the applicant, was waterpipe."

Did we need to classify it as one or the other?

Were there *qualitative* differences as well?

And what were those quantitative differences he was *imagining*?

In other instances, we'd clarify contradictions like this. We'd reach out to the applicant or consider it also as pipe tobacco. In this case, what was different? We hadn't finished these reviews, so why had he said "the review *was done*"? We were still deciding how we'd approach these reviews.

And what did these people's job descriptions look like? Instead of considering facts, decisions were made based on unsubstantiated *beliefs* or on just what someone *thinks* at the time, without any justification. What he *imagined* would preclude *any* consideration of qualitative or quantitative differences in emissions.

Some colleagues and leadership thought the industry was trying to sneak one past the goalie. By including two product categories on the packaging, they'd expand their market beyond the marketing order they'd pay for and receive for only waterpipe tobacco. That could certainly be the case, but there might've been a less sinister explanation.

While at the tobacco shop, I'd noticed other waterpipe tobacco packaging said things like, "TTB CLASSIFICATIONS: Pipe Tobacco." So, I checked the TTB website. The U.S. Department of the Treasury's Alcohol and Tobacco Tax and Trade Bureau (TTB) required a permit to import or manufacture a tobacco product in the U.S. There were specific permit categories: cigarettes, cigars, chewing tobacco, snuff, pipe tobacco, and roll your own tobacco. There was no category for waterpipe or hookah tobacco. So, it looked like the confusion might've been on account of industry's compliance with another federal agency. TTB had no waterpipe category, possibly explaining "Ingredients: Pipe Tobacco" on the waterpipe products we were considering.

I shared this info with my leadership and reached out to TTB to confirm no waterpipe category existed. This seemed like an interagency opportunity to clarify labeling in a way that would benefit the public health. CTP never responded, and TTB only responded with requests to fill out a survey about my customer service experience. Oh well.

On July 17th, 2024, CTP then issued a determination that we'd disregard labeling as "pipe tobacco" because the font size was small, because "pipe tobacco" was probably shorthand for "waterpipe tobacco," because a change or modification to the label was outside CTP's regulatory authority, and because previous reviews did not raise this concern.

Competent staff involved with that determination felt demoralized. The issue under consideration didn't involve a change or modification to the label. That determination said:

Social Science [Branch] *reviewed and considered that there is zero to very small likelihood that the labeling would be misunderstood by a consumer. The statement about this being "pipe tobacco" is in very small print and is couched within other statements that are indicative of waterpipe tobacco. Overall, Social Science did not have concerns.*

That wasn't true. "Ingredients: Pipe Tobacco" stood out on its own, and it wasn't "very small." And that statement, before it was edited, was shared with me:

Social Science [Branch] *reviewed and replied that they believe there it is a zero to very small likelihood that the labeling would be misunderstood by a consumer. The particular statement about this being "pipe tobacco" is in very small print and is couched within other statements that are indicative of waterpipe tobacco. Social Science does not have concerns but agree that if there was misunderstanding there would be different exposures, and different considerations as the populations who use the two products are dissimilar.*

CTP provided no explanation of how a "zero to very small likelihood" had been determined. They wrote about "pipe tobacco" on the label:

This could be understood as a shorthand (waterpipe shortened to pipe).

So, accurate product labeling is apparently no longer important. There might be additional ingredients in products besides those listed on a product label. And beware those with peanut allergies when you see "butter" on a label:

This could be understood as shorthand (peanut butter shortened to butter).

And the last bit of the determination was telling:

Similar SE Reports [applications for substantial equivalence] *have the same labeling, but the SE reviews did not include deficiencies or Environmental Science requests.*

Like my Secondary Reviewer #2 said, "We just happened to catch it when no one else did." And like before, I later learned the same or similar labeling discrepancy had been missed during previous reviews. No one bothered telling me that. We couldn't do things differently now. So, instead, they came up with one ridiculous excuse after another.

Again, changing gears would risk litigation that would undermine the good work we did, and that might also undermine the good work of other FDA centers. Over time, however, those reasons to support the status quo looked more like a means for complicit managers to hide things. And how *good* was our work? If they really cared about our public health mission, they'd correct mistakes and allow more credible work going forward. These tactics seemed more like self-preservation at the public health's expense.

26 | "It melts in the mouth."

IN MID-JULY OF 2024, I received a new assignment to review 20 oral pouch products via the Premarket Tobacco Product Application (PMTA) pathway ideally completed before a product goes on the market. Those 20 products deliver nicotine to users when placed between the lip and gum. In 2020, the manufacturer submitted to CTP applications for marketing orders for those products. However, they were already sold in the United States. CTP allowed the applicant to sell those products in the U.S. for years while the agency made its determination about if they should.

Interestingly, our Office of Science Director #1 left CTP in July of 2022 to work for the tobacco company that acquired, in late 2022 and early 2023, the manufacturer of those 20 products. So, he represented the regulatory agency—then the applicant—while those 20 products were reviewed.

FDA and CTP leadership had just been grilled in front of Congress about the agency's failure to meet its regulatory deadlines, for allowing products to be sold without marketing orders, and about delays granting marketing orders for products that might be lower risk. Sales of those 20 products were skyrocketing. However, CTP hadn't yet determined if oral pouches, like those, marketed by *any* manufacturer were appropriate for the protection of the public health. Instead of containing cut or ground tobacco, like in smokeless products or snus, these 20 products instead contained nicotine *extracted from* tobacco.

The company had already responded to a group of requests from CTP about those 20 products when I received the assignment. CTP tobacco product reviews could involve more than one cycle. In general, a cycle constituted CTP's review of product information and either a determination, where a marketing order was denied or granted, or questions the applicant could answer if they wished to continue to a subsequent cycle where their responses would be considered. Otherwise, they'd not receive a marketing order.

In this case, unlike the previous examples where my work started at the beginning of the first cycle, I was coming on board halfway through the second cycle. The first cycle review had been completed by a different primary reviewer with a team of secondary and tertiary reviewers, other primary, secondary, and tertiary reviewers from a variety of other scientific disciplines, a Technical Project Lead (who collectively considered the science), and a Regulatory Health Project Manager (who managed the review process). Each cycle's reviews and questions to the applicant were further scrutinized by those with expertise in science policy and by CTP lawyers.

Reasons for a second round of reviews varied. In this case, CTP needed more information. What the applicant initially provided hadn't been sufficient to prepare the 20 assessments. It was not odd to ask questions and conduct a second cycle review, and it appeared when I did a preliminary review of those responses that the applicant had provided the information that CTP had requested during the first cycle.

My job was to determine if the information provided by the applicant and otherwise available about each product seemed sufficient to perform an assessment of each product. The review that I would perform would communicate that outcome. If so, I would subsequently prepare one programmatic assessment that would include evaluations of all 20 products. Our reviews only communicated the adequacy of the available information to perform an assessment, not the outcome of an assessment. The outcome couldn't be determined until the assessment had been completed.

July 22nd, 2024

At our first meeting about those products, my Secondary Reviewer #3 said that CTP had decided to grant marketing orders for 14 of those 20 products, and therefore, CTP needed assessments for 14 of those products. He said CTP likely wouldn't let us raise any issues or ask the applicant any questions, but he said that I'd still need to raise any issues that needed to be raised.

How would any of that matter if the decision to grant those 14 marketing orders had already been made? The outcome of the assessment couldn't be determined until the assessment had been completed, but CTP determined the outcome beforehand.

August 1st, 2024

Our Supervisor #6 joined us for our second meeting, where they discussed concerns about microplastics associated with those products. They discussed my ability to tackle those issues, considering my expertise at CTP with microplastics. Supervisor #6 said, "So, I don't know if you guys saw the research priorities, but they specifically mentioned microplastics." She said to me, "Yeah, this would be a great opportunity because you're involved in the microplastics working group." She mentioned the Microplastics Exposures from Tobacco Products Project I'd led since 2021, and she said microplastics were a new research priority for the Office of Science.

My Secondary Reviewer #3 said the products constituted a "pouch with microplastics, so, that is the one that Christy would be great to sort of dive into if there are facets of it that we need to raise." I therefore mentioned the concern about users of those products consuming microplastics and described the microplastics initiatives I'd been involved with over the previous years.

They talked about doing the assessment for those 14 products in the context of what was currently known, to reflect the best, most up-to-date information.

August 2nd, 2024

My Secondary Reviewer #3 and I had what seemed like a good discussion focused on doing a credible review. He'd prepared a draft assessment during the first cycle that we could update but talked instead about using the applicant's assessments. Doing that would mean we wouldn't add anything, and the public would only see unredacted parts of what the applicant provided about those products.

We talked again about leadership's concerns about microplastics. I told him, "I've been sharing [microplastics information] with my chain up to [our Office of Science Director #2]."

He said the pouch was "dissolvable."

Other work was then prioritized, and we each took several weeks of leave.

August 27th, 2024

During a training for our branch of about 25 scientists tasked with CTP's NEPA compliance, the same official messaging I'd heard before was repeated. Branch Chief #3 and #5 previously said we didn't consider product use. However, the trainer again said that we were to "quantitatively assess the impacts of manufacturing, use, and disposal" and he pointed to the importance of marketing estimates in that context.

My Secondary Reviewer #3, Supervisor #7, Branch Chief #3 and #5, and Branch Chief #7 attended. They each spoke or asked clarifying questions. So, they heard what I heard again that day. The trainer said, "So, communicating with the primary reviewer, specifically as a secondary reviewer, when we're reviewing documents and making comments, and things like that, we should avoid phrases like, "you, please, should, need to."

September 27th, 2024

I asked about a timeline or any deadlines for our work on those 14 oral pouch products, but there were no deadlines and there was no timeline. My Secondary Reviewer #3 said, "They might not act on this until after the election." He said Office of Science leadership informed Regulatory Health Project Managers, who coordinated the logistics of our review process, that "it's not going to happen until mid-October, at least."

October 4th, 2024

Branch Chief #7 called an impromptu meeting to discuss the applicant's assessments. My Secondary Reviewer #3 said, "If we use the applicant's assessments, we would actually give more information to the public." Although we could make our assessments more informative, he said, "When we do our assessment, it's not as informative."

Their conversation then implied we'd shift gears and write our own. That involved a very different process. So, I said, "I need to think about this from a primary reviewer's perspective, because we haven't talked about writing the assessment. That's a different process."

Higher-ups determined if applicant assessments could be used, and that's what we'd previously talked about. Me, instead, doing that meant those assessments needed to be carefully prepared. These were the first oral pouches of their kind CTP would grant marketing orders for, so there was no product available to serve as a comparator.

Branch Chief #7 said the decision was up to us. That was fine with me because it meant we could better inform decision makers and the public. She said she didn't like the formatting in the applicant's assessments, then pointed out other things she didn't like. On the screen, they quickly scrolled through documents, pointing to things they agreed or disagreed with, and elements that should or shouldn't be included, and then we'd be done.

So, I said, "As the primary reviewer, I need to get my head into this draft assessment. I don't want to say, 'I'm comfortable with this,' scrolling very quickly on the screen, as a primary reviewer, when I haven't had the chance to really consider that." I said, "I need to think about it."

My Secondary Reviewer #3 agreed, "Fair enough. That is fair."

I said, "We need to make sure the pieces are there. We need to make sure it's accurate."

Branch Chief #7 became argumentative, as if we'd hindered the process. We hadn't. We'd reached out to her and other appropriate contacts to coordinate our efforts, and we'd gotten crickets for weeks. Her behavior contradicted our training and what higher-ups said about our process, then, and recently when—instead of doing accurate work—we'd been allowed to remove statements attesting to the accuracy of our work. We were working around doing credible work, and we needed to get our purview straight. They were telling me what I did was my choice while they were telling me what to do. So, I said, "If I'm going to jump in and put my name with my credentials on a PMTA, I've got to understand more about our process, I don't—"

My Secondary Reviewer #3 then typed in the chat that he had another meeting and had to go.

Branch Chief #7 said, "I do too."

I quickly asked, "Do I need to agree with the argument in the assessment and do my background and my credentials need to correspond with the information in the assessment?"

My Secondary Reviewer #3 said, "Take your time with this, as you need. Please take your time on that."

I responded, "I appreciate that," and again asked, "Am I expected to agree with the argument?"

Channeling her inner Nathan Thurm, Branch Chief #7 then said, "I can't say that I understand you when you ask that question."

They then talked about other things. So, I typed in the chat that I'd move the question to an email.

She snapped, "No! That will not be necessary, Christy!"

They then talked again for some time about other things. So, I tried again, "I really would appreciate if, I just need some documentation about what my role is as a preparer and if I need to agree with the argument. I think we need to move this to an email as well because we need documentation of this."

Branch Chief #7 snapped again, "No Christy! I called this meeting so we won't have to send emails back and forth!"

I asked, "[Branch Chief #7], are you saying I cannot document this question?"

She then backpedaled and snarled, "You can. You can do anything you want, Christy. I can't stop you."

Weird behavior, but good to know. I moved the question to an email, but she still refused to cooperate. So, I had to do as our Division Associate Director, Deputy Director for Regulatory Management, and Office of Science Director #2 directed. The arguments in those assessments would need to be mine—they'd needed to align with my oath, training, credentials, and experience. I'd need to raise what needed to be raised. And we'd need to use the best, most up-to-date information.

October 7th, 2024

The next time I met with my Secondary Reviewer #3, he concluded, "There is actually no real impact from use [of these products] because it's a dissolvable pouch, and probably excretion after use is one part of it which, could be the microplastics and I don't know, I don't know, I don't know about that. And for disposal, it is pretty much the, you know, the packaging material that is the only disposal, unless we want to talk anything about excretion."

His messaging to do credible work now shifted into discussions where we'd identify a selection of information that would facilitate marketing orders, before the review and assessment had even been done.

"I am an environmental scientist," I said, and "I understand that if we just start cherry-picking information to show there's no environmental impact, that's not good. And that's what we're talking about right now." I said, "What we're doing here is like, 'Oh, we can find a study that shows this is not a problem.' This is unethical, right? Our job here isn't to find information to prove this is not a problem in the environment, it's to do an assessment. Again, that's my understanding until I've been told in writing that 'No, you're supposed to cherry-pick to show that there's not an effect, so this can go on the market.'" I said, "This is a real problem."

Then he, too, voiced concerns. He said, "At many levels, they are just using our scientific integrity to get through shady documentation [from the applicant], is what I feel about us writing the assessments." He again said, "There is no urgency here. They are not going to issue marketing orders for [product name] before the election."

October 8th, 2024

My Secondary Reviewer #3 then said, in lieu of ensuring clear and documented directives, that he did a "reading of the tea leaves" about what our Branch Chief #7 wanted. Branch Chief #7 now consistently followed suit with her predecessors. She provided contradictory messaging that feigned credible work while avoiding any engagement with concerns about cherry-picking. Instead of doing assessments that included what we knew and that informed decision makers and the public, he said that she asked if I could simply "pull" information from what the applicant provided, cut and paste certain bits and attach my name, credentials, and experience, as if it was my work.

I said, "She needs to put it in an email. I don't read tea leaves for the FDA."

October 10th, 2024

The next time we talked, my Secondary Reviewer #3 again told me, "There is no real use impact for this product because people don't even spit anything out. Unless you want to be really picky in the context of excretion, right, whatever comes out in the form of excretion. That would be disposal, I guess."

I asked, "Can you remind me what the pouch is made of again?"

He responded, "It is a pouch that is tobacco-free and dissolvable cellulose," but it wasn't. Referring to it as "tobacco-free" was a misleading characterization some staff were concerned about. The nicotine in these products came from tobacco and therefore likely contained the same carcinogens—tobacco-specific nitrosamines—found in many other tobacco products. And all evidence indicated that the pouch did not dissolve.

I asked again, "So, the pouch itself doesn't persist either?"

He responded, "No, it melts in the mouth."

October 16th, 2024

At our next meeting, we talked about how this assignment was handled differently. He again said they wouldn't "authorize it before the election," as if that provided an explanation.

In the past, we'd have meetings with the Regulatory Health Project Manager that coordinated our efforts. We'd meet with the Technical Project Lead that pulled together all the reviews from different scientific disciplines and collectively considered what we all said. We'd all meet to share information across disciplines that might be important. And a list of dates for meetings, milestones, and when draft documents should be provided, reviewed, and discussed at each level, before they were finalized, had been provided for every other assignment. Not this time.

I'd worked on PMTA applications that CTP had denied marketing orders for. In those cases, assessments weren't provided, only a review. But in those cases, those weekly meetings kept us on track and ensured we did all we could to substantiate those denials. In my experience, we did better work when CTP denied marketing orders.

This time, we had none of those meetings and our work was markedly less comprehensive. We grappled with how to do credible work while pressured to cherry-pick. Our discussions increasingly vacillated between doing good science and supporting the status quo.

"I know that we don't have benchmarks," he said and asked, "But, based on your judgment, do you believe there is a significant environmental impact from this product from disposal?" We were frequently asked what we *believed*, not as scientists, but otherwise as if that should trump a scientific approach. And he was asking only about disposal. What about use?

"I don't know. If you're asking me as a scientist, that's my answer," and I asked, "Are you asking me as a scientist or you're asking me as a person that's just making a guess?"

He said, "Both."

I said, "Well, in what capacity am I doing my job? So, I think that matters, right?"

He said, "Okay."

I said, "If I'm hired as just some person off the street, that's not being a scientist, that's the answer people want me to give, right? So, I understand what it takes to make a scientific conclusion." We hadn't yet done the assessment, so we didn't know. I continued, "I don't need to agree that I have done a proper scientific evaluation of [product name] disposal because we haven't done that. We will not do that. That is not what we do. We cherry-pick."

I asked, "Do I *think* there's going to be an impact? Think about if would you want the public to hear what you're saying? And this goes across anything, when I'm having an internal discussion at FDA about a product and I'm asked, 'Do you *think*?' I'm *thinking* as a scientist. I have to. It's my job, right?"

He responded, "Yeah."

"I've got to stick with, like what my job *is supposed to be* to some extent," I said, and "I've also, ethically, just got to do the best that I can." I said, "I just want to make sure that when we talk about these things, I can't, we are cherry-picking and I want to be really careful about just looking around for information that supports the FONSI."

"I agree with you," he responded, "But, at the same time, you know, we should be able to connect the dots in terms of, and I am also on the same page with you that it's not for the FONSI, but I don't see any compelling glaring thing that would point at, 'Hey, this might pose a problem for this product.'"

We hadn't completed the assessment, so I said, "Yeah, so based on my knowledge, I agree with you," but, I said, "It starts making it sound like we're trying to convince people [including me] of something, and I'm concerned about that." Why did he need me to profess believing something before we'd even completed the review and assessment?

During what initially seemed like a more objective part of the conversation, we talked about constituents of the products and that users consume the products. That was unique for products regulated by CTP. Constituents of those products would move through or become integrated in biological systems during and following use. This was an odd, but nonetheless valid, means of disposal for constituents and their breakdown products, like plastic polymers.

He said information "we have available for GRAS is valid in this case because it is an oral pouch." GRAS refers to something generally recognized as safe, usually in the context of food, and he seemed to grasp at straws to dismiss consumption before we'd even talked about it.

I recalled a discussion at a branch meeting and a pending memo that would ensure we applied any GRAS designation appropriately. I said, "I remember [our Branch Chief #3 and #5] saying at one point that products [and constituents] designated as GRAS for food did not apply in tobacco, for tobacco products."

He said, "That is mainly because of the route of exposure difference between tobacco products [and food] in general."

I said, "Right. Right. But, in this case, this is consumption." We'd need to confirm that, and it didn't apply, but he later used GRAS as a means to dismiss concerns. Because something was GRAS in a food product did not make it GRAS in a tobacco product.

He said in the context of air emissions, "We could say while these chemicals are found in different quantities, these are all absorbed, right, because these are all ingested, and excretion is the way that this [any contaminants] goes out."

What?

It struck me as odd that we'd present the human body as a filter that would prevent consumed constituents of tobacco products from polluting the air. So, I summed it up sarcastically with a tinge of disgust, "You could say there are no air emissions because the liver and the kidneys and the rest of the body is going to absorb all this shit."

When there was a potential concern about the *use* of products, where constituents contaminate the bodies of users, we would instead focus on *disposal*. And we'd only consider *disposal* in the context of *air emissions*, because if we wanted to find no impacts, we needed to look where it was the least likely to see any.

Not acceptable, so I said, "I think we should request something from [our division's] Toxicology [Branches], whether it's as a confidential appendix, something," because they should've been considering that sort of stuff.

He responded, "I don't think anybody's going to stop you, nobody's going to stop us from doing that."

Our Division Director had newly written that any request like that should go through our Supervisor #7, who'd recently replaced Supervisor #6. So, I asked Supervisor #7 about doing that, but she wouldn't respond. I, therefore, asked Branch Chief #7, but she wouldn't respond. I eventually let our Division Director, who wouldn't respond, and therefore also our Office of Science Director #2, know that they were unresponsive. They then said to reach out again to our Supervisor #7, who again wouldn't respond. So, they then said to reach out to our Branch Chief #7, who again wouldn't respond. That behavior persisted until I left the FDA.

My Secondary Reviewer #3 said we still didn't have a deadline or timelines, because "they're not going to approve this before the election," and other work was intermittently prioritized.

October 25th, 2024

During our next meeting, we again noted it was odd to still have no timeline or meetings. He again said, "They're not going to do anything on this at least until after the election," so he said there was no rush.

Still, we thought we should prepare. We substantiated that arguments used to indicate no impacts in other assessments didn't apply in this case. For example, we talked about how we had no comparator product and sales were increasing. We couldn't write these products off as new products entering a declining market where therefore there'd be no additional impacts.

He talked about how he'd chatted with a colleague who worked for CVM, FDA's Center for Veterinary Medicine. He said, "He talked about how they do [only] five to six assessments a year," while we cranked them out like an assembly line: cut, paste, sign, and stamp. He said the interaction between industry and CVM was vastly different. He said, "They can ask for studies to be done." He said, "Apparently they can request all sorts of information from the applicant."

We talked about possible reasons CTP leadership was reticent to do credible work. We talked about tradeoffs, how we were expected to support the sausage-making because the long game would be worth it, where we could do things like limit the amount of nicotine in cigarettes. But why did that necessitate a tradeoff? And we didn't have the authority to orchestrate those tradeoffs.

It seemed while enforcement discretion should be explicit, it instead was used willy nilly with abandon.

And how long was the long game?

What about the new addicts, death, and disease we'd be responsible for in the interim?

We wondered why CTP used dated approaches lagging 15 and 20 years behind current science applied in regulatory contexts elsewhere. We talked about the frequently referenced influence of litigation and about personalities unwilling to admit mistakes. Some seemed so thin-skinned that they might sideline the public health to attend to their egos. But tactics were too consistent to be the outcome of uncoordinated individual actors.

I'd recently visited FDA's White Oak headquarters. I told him colleagues from other FDA centers reacted when noticing "CTP" on my name tag, "'Oh, you're from CTP,' and then they just started telling me shit, man, about the reputation of the center. Other FDA centers believe that our center is corrupt and they're looking at me like I work for the industry. Some were very sympathetic and they'd say, 'How long have you worked there?' and 'Okay, here's some advice about what to do and not do.' And I'm like, 'Too late. I already did it.'"

Their advice was to not raise concerns, but that was my job.

We talked about our ability to source information for our assessments. Our research projects were designed explicitly to inform our assessments. He said, "Nobody's going to stop us from having this cross-pollination at a local level," with respect to our ability to do *research* with other scientific disciplines, and he continued, "But where the dilemma or where the challenge lies is actually translating it, to putting things into our work." We knew so much that wasn't reflected in our *regulatory work*.

I said, "I don't think these [products] are dissolvable completely," but he didn't respond. And the more we learned, the less likely the assessments could reach findings of no significant impact, especially considering the expectation that we mirror past rationales that didn't apply.

October 28th, 29th, and 30th, 2024

I attended the American Public Health Association 2024 Annual Meeting & Expo, "Building Trust in Public Health and Science," in Minneapolis, Minnesota, and the timing couldn't have been better.

CTP's delayed decision making and inaction to prevent the sale of unauthorized products, oddly, sometimes benefited us. Unwitting American guinea pigs provided data about products that hadn't yet passed through CTP's regulatory review, but that nonetheless were sold and used in the U.S. And we were supposed to take any credible information into account, not just what was provided by applicants hoping to sell products or what compromised CTP staff pushed. And what was confirmed didn't require teams of university researchers.

I'd just done my review of information about those 14 oral pouch products that were discussed at that meeting by many from a variety of perspectives. By happenstance, I sat next to a researcher from The Ohio State University who saw "CTP" on my name tag and said she was involved with research about the products I was reviewing. It was interesting to hear so many discuss the very same oral pouch products that I was reviewing. Everyone else, everywhere besides CTP, knew those pouches didn't dissolve.

I had an idea. It would harder for CTP to pretend those pouches dissolved, or ignore me when I said they didn't, if I had those pouches in hand. So, after sessions ended on the 28th, I drove to the Mall of America. I walked to the Hideaway on the third floor. I recalled seeing smoking paraphernalia in the window and thought they'd carry those products. However, a sign in the window said they didn't sell nicotine products. When I asked a store employee where I could buy those products, he said, "You'll need to go to the street corner."

What? I said, "What?" Although those products hadn't yet been granted marketing orders, retailers didn't care. I thought I'd seen them sold everywhere. Why would I need to go to the street corner?

He said, "Street Corner News, on the first floor," a store on the first floor.

And there they were, behind the cashier, one can of 15 pouches for $14. At 6:19 pm, I snapped a photo of the can and texted it to my Secondary Reviewer #3. When I got home, I stuck a pouch in my mouth, in the microwave, in water, in water in the microwave, in a cup with spit, and in a cup with spit in the microwave. I waited and I watched. I sucked them and soaked them and smashed them, and of course they didn't dissolve.

And our process still wasn't written, but Supervisor #7 had finally pointed to a loose group of resources that should be referenced while implementing that unwritten process. My Secondary Reviewer #3 saw the same resources that indicated we should follow the agency's recently updated scientific integrity policy. What our Supervisor #7 pointed to confirmed we couldn't reach findings of no significant impact with the information that we had.

So, I let my Secondary Reviewer #3 know, "The [product name] assessments will therefore not support findings of no significant impact. Please let appropriate leadership contacts know. I'm happy to provide any needed information for discussions. Please let me know about any such requests so I can prepare."

With around 40 CTP employees and contractors, our CTP Director #2 also attended that American Public Health Association meeting where discussion of these products was front and center. I served in an "Ask the Expert" session in our CTP booth with him. Attendees asked about CTP's efforts to address secondhand emissions. He said that was outside of our regulatory authority, but it wasn't. He also lamented about his inability to unplug during his upcoming honeymoon in Bora Bora. He said he might get an urgent work call about regulatory work on hold that might soon break free and require quick action.

November 1st, 2024

During our next meeting, my Secondary Reviewer #3 asked why the assessments wouldn't support findings of no significant impact. Most of those reasons he'd raised, so he agreed with the rationale.

These would be the first products of their kind to receive marketing orders, so we couldn't say they replaced other products in the same category. We had no basis of comparison and couldn't say anything was similar to products already legally on the market. We couldn't say the products were part of a market in decline because the market was skyrocketing. And we didn't know what the pouches that users put in their mouths were made of.

Unlike every other assessment I'd seen, we had no comparator product. Our Division Director had just written that we didn't need to include in our assessments harms associated with a new product if those harms didn't differ from harms for the collective of products making up the class of products to which the new product belonged. CTP categorized products in nine classes: cigar, cigarette, ENDS (electronic nicotine delivery system), HTP (heated tobacco product), pipe, roll-your-own, smokeless, waterpipe/hookah, and *other*.

We were told that our comparator couldn't be smokeless pouches, which include as ingredients cut, ground, or powdered tobacco, or tobacco leaf. The products we were considering instead contained nicotine extracted from tobacco, with other ingredients, for example, like sweeteners, flavors, preservatives, and fillers. So, these 14 products would be the first of their kind to receive marketing orders categorized as *other*. I found in our online Searchable Tobacco Products Database, where all products receiving marketing orders should be listed, only four *other* products: VERVE Chews Blue Mint, VERVE Discs Blue Mint, VERVE Discs Green Mint, and VERVE Chews Green Mint marketed by U.S. Smokeless Tobacco Company LLC. Those were chewing gums and couldn't serve as comparators.

We had no data or set of considerations associated with the class of products to which these new products belonged and could be compared. Regardless of how appropriate it was to exclude from an assessment information about a new product solely because that product was from a class of products with likely similar harms, in this case we had no basis to exclude anything. And it certainly did make sense for the first marketing orders for a new type of product by CTP, in its role to do that only when those actions were appropriate for the protection of the public health, to be informed and inform the public.

Likewise, the rationale used to demonstrate marketing estimates informed the agency's decision making didn't apply here. Supervisors #2 and #4, our Division Director, and others said the agency determined there were no impacts because the same number or fewer units of a given new product would replace a comparator product that would be removed from the market. Without a comparator product and considering these products constituted the first of their kind granted marketing orders, every unit that went on the market constituted an increase in any known, potential, or unknown harm associated with their manufacture, use, and disposal. Our Division Director had recently written "I do not think that we have encountered such a situation yet," "some kind of specific harm to humans that is different from products of the same class." Now we were there, where any sort of harm would be new for this class, and we didn't even know what the pouch was made of.

Documents previously prepared by other staff indicated that information about the materials constituting each product and its packaging were known but nowhere were the materials constituting the pouch described or discussed, not even in a confidential appendix. Potential problems associated with the use and disposal of the pouches were likewise omitted. Additionally, we still didn't confirm marketing estimates that were provided by applicants. And these marketing estimates were provided in 2020, and we hadn't requested updated estimates. So, even if marketing estimates had been accurate in 2020, they wouldn't be now. Our training directed us to consider other information besides what the applicant had provided. So, we knew from online, publicly available information and from CTP's tracking of these very products that sales had increased quite a bit. We'd also been told that we needed to verify how those marketing estimates had been determined.

CTP's decision makers should've noticed that dated information and those omissions, and they should've wanted these things properly considered.

For several years, I'd served as CTP's subject matter expert and representative on various initiatives related to plastic polymers, nanomaterials, and microplastics. In 2021, I'd developed and since led a project summarizing potential health and environmental concerns associated with plastics in tobacco products. That project proposal included a summary of published research indicating potential concerns at every level of biological organization. My Supervisor #2, all our branch leadership, our division, and the Office of Science approved that project proposal. I shared project updates with them, with all staff in my branch, and with CTP's Office of the Center Director for which I also reviewed pending treaties and other legislation related to plastic polymers and microplastics. CTP leadership was therefore aware of the body of scientific evidence amassing about plastics and microplastics. My Supervisor #2 was even an active participant in that project for over two years.

In July of 2024, I'd also shared with them FDA's new Microplastics and Nanoplastics in Food webpage. The agency focused on food as possibly the most far-reaching route of exposure across all the products the FDA regulated. However, exposures happen from products regulated by all FDA centers.

Smokeless and other nicotine pouches were a primary focus of the project I led. It later expanded to include cigarettes, cigars, and ENDS products, but oral pouch products were the initial focus—because the agency knew about increasing concerns associated with plastic polymers and exposure to microplastics. The explicit focus of that project was to inform our assessments. The approved project proposal and project updates consistently communicated potential concerns with exposure to microplastics released during and following use of oral pouch products. For example, we needed to understand how working the pouches to release their contents or simply holding a pouch in a warm mouth full of digestive enzymes influenced the consumption of materials constituting the pouch. Until we knew, we needed to at least include mention in our assessments of possible harms and uncertainties associated with potential exposures.

Oral pouch products made of plastic polymers will degrade into small plastic particles. So, users might eat microplastics when they use oral pouch products that are made of plastic. Some percentage of those microplastics will likely be excreted in their feces and therefore end up in septic and sewage treatment systems along with pouches that are flushed, something we hear youth especially do to hide their habit. Flushed pouches can be problematic in septic and sewage treatment systems and will also release microplastics. Further, microplastics will be released where plastic pouches are littered, for instance on roadsides, on beaches, and in streams, and where disposed of properly in municipal solid waste.

Microplastic pollution is difficult to contain and will subsequently end up in biological systems, in us, via ingestion of contaminated food and water and the air we breathe. But what might be the weirdest means of *disposal* is the direct incorporation of microplastics in the tissues of oral pouch users, in the smokeless and *other* product categories.

All that said, these considerations of use and disposal assume pouch materials align with what is known from patents and from other publicly available information. We couldn't assume any of that applied until we knew what those 14 products were made of. Users could be putting something entirely different in their mouths. The actual materials could pose more, or less, of a risk.

However, my Secondary Reviewer #3 now said we couldn't ask the applicant for information. We knew the ingredients held *within* the pouch that users put in their mouths, but not what the pouch *itself* was made of. That information hadn't been provided. We could've asked. We could've discussed that information in general terms in the assessment and listed the materials in a confidential appendix, respecting any confidential business information.

We didn't need to assume things we didn't yet know, but we needed to at least throw a bone to potential exposures. We needed to adequately flesh out what users might be exposed to during use. We needed to adequately consider disposal after use in municipal waste or as litter. Disposal in the tissues of users should've also been explored as best we could and communicated to decision makers and the public, along with any uncertainties. But fundamentally, not knowing the materials that constituted the pouch was a dealbreaker.

Online information about products with the same names clearly indicated the pouch didn't dissolve. And while the pouch materials weren't described in application materials, instructions included there about their use indicated the pouches didn't dissolve. On the bottom of the can I bought it says, "Pouch and contents are not intended to be consumed. Discard after use."

It was strange that CTP claimed ignorance. So, I said to my Secondary Reviewer #3 at that November 1st, 2024, meeting, "It's all over the internet. It's been there for years. So, we are not doing our due diligence in the branch. And I need to know why, because, you know, there are these efforts, these discussions I have with my colleagues and we're like, 'We need to do this scientifically.' Well, why aren't we?" I said, "I will raise issues. So, I just want to make it clear because I think a lot of us think that we're helping when we're actually just loading up the assessment with information that makes it look like we're doing something scientific. Do you get what I'm saying?"

He agreed, "Yeah, no, I know."

He said excluding the pouch materials had been an oversight. He said they'd erroneously excluded any considerations of the pouch materials, thinking the pouch dissolved completely. For the over four years before I joined the assignment, apparently, no one noticed. A complete round of reviews had been completed under the assumption that the pouch dissolved. Those involved before me during the second round also, apparently, made the same mistake. However, in meetings about these products just a few months earlier, we'd talked about concerns with users ingesting microplastics associated with these very products.

We talked about similarities between these pouches and cigarette butts, the most littered item on the planet that, for some reason, our branch *had* considered and was very interested in tackling. Why consider that and not oral pouches? He referred to that as the responsibility of the U.S. Environmental Protection Agency. But if we thought that, we should've consulted them and provided that information in our assessments. He said FDA would grant the marketing orders, and then it becomes "someone else's problem." Considering that rationale, why had we dedicated resources to research about cigarette butts?

I said, "I listen to [our CTP Director #2] say it over and over again, 'We use a robust, comprehensive scientific approach at CTP.'"

He replied, "No," in agreement, that we didn't.

On June 12th, 2024, our CTP Director #2 had provided testimony during a hearing before the U.S. Senate Committee on the Judiciary: "Combatting the youth vaping epidemic by enhancing enforcement against illegal e-cigarettes."

He said, "We have a premarket paradigm to review the scientific evidence, to understand what is in the product itself and to make a determination based on the science, whether they meet the appropriate statute by law."

He said, "As a scientist myself, I'm committed to following the science, and when I took this job, I vowed that and I'm continuing it now," and then confirmed, "We have an important paradigm in place."

When asked, "What are you doing to mitigate the harms that Juul is causing while you complete your review?" he said, "Well, part of it is the scientific review, to make sure that we're doing a robust review of the evidence, including the toxicologic risks and various other risks at the population as a whole."

Why would Juul be subjected to a robust review, when other products weren't?

He said, in reference to Elf Bar, "We know Elf Bar and those progeny products are the most commonly used product among kids, and that's where we prioritize the resources," and "so it's important that we follow the science."

The oral pouch products I was reviewing were popular among kids, so it was important that we take the time needed to follow the science. That should be important regardless. But during that June 12th, 2024, hearing, Congress expressed frustration about delays in marketing orders for non-combustible products.

External messaging constantly contradicted what we did. For example, likewise, our Deputy Director for Regulatory Management said at that October 24th, 2023, public meeting, "Premarket Applications: Opportunities for Stakeholder Engagement," that "Every individual product has to be assessed for its individual combination of risks and benefits. That's required by the law."

He said, "Making a presumption is something that we just aren't comfortable doing, even when there's a lot of data. We don't do that with cigarettes, and we've had a lot of data on cigarettes."

He said, "And so even though you may say a product is only the ingredients, we need to look for the issues if we're still looking for other things that are cropping up that we have be aware of."

In this case, we weren't even considering all the ingredients. We weren't considering other things that were cropping up in this group of 14 PMTA applications, like a skyrocketing market. We might've even been presuming and assuming our way into breaking the law.

At that same public meeting, our Division Director said, "Right now tobacco products carry risk. And the PMTA rule, the ENDS guidance, the Tobacco Control Act—they require—well, the law requires, the rule requires that we describe, fully describe the harm or the toxicological risk from the product."

Then, on September 10th, 2024, our CTP Director #2 testified again, this time to the U.S. House of Representatives, Committee on Energy and Commerce, Subcommittee on Health in a hearing, "Evaluating FDA Human Foods and Tobacco Programs." Likewise, when asked, "Why hasn't FDA issued a marketing order for a single synthetic nicotine product?" he said, "We haven't issued an authorization because no applicant has provided sufficient evidence to meet the necessary scientific bar."

He said, "As a scientist, I think that a scientific bar is critical, and I think Congress got it right in requiring that scientific standard to make sure that the products we are authorizing are meeting the appropriate for protection heath, protection of public health standard that Congress [intended]."

He said, "We are required by law to review the individual merits of the application."

He said to protect public health, we have "a scientific standard and evidence used to determine what should be authorized."

So, during that November 1st, 2024, meeting with my Secondary Reviewer #3, I'd continued, "[Our CTP Director #2] keeps saying it, so, um, yeah."

My Secondary Reviewer #3 responded, "It's not all about science! How do you say that out loud?!"

I said, "Everybody knows that. The problem is he keeps saying it. And if he doesn't want me to do it, he needs to stop saying it. I am not here to just help market products. That is not my job description. My job is to do a scientific evaluation."

He agreed, "Yeah."

He thanked me for raising the issue, apologized for the misunderstanding, and said, "I'll own it, I thought it dissolved. I did think that. I should have known better. I didn't realize that it does not dissolve." But it wasn't just him. So many others looked at the same information. He said, "Thank you for bringing it up, you're right." He said, "We could write a review that points to these [concerns] and ask them to address it." He said, "We can ask the applicant a question."

Then at the end of the call, I asked, "You haven't heard anything from anybody about a timeline or anything? Nothing?"

He said, "No, nothing, nothing, nothing, right?" And he said again, "So, I do, I appreciate you, um, you know, opening my eyes to some of these problems. I misunderstood."

November 7th, 2024

During our next meeting, my Secondary Reviewer #3 now vacillated histrionically between credible work and the status quo. As his radicalization took hold, he became disorganized and emotional.

He started out saying, "So, what, um, one thing that I wanted to tell you was that I actually was at a meeting with, um, [our Supervisor #7] and uh [our Branch Chief #7] about uh [the product name] and they, uh, for a lack of a better way to put it, they want me to work with you and wrap this up. Apparently, it's with the Office of Compliance and Enforcement for final clearance and to the extent that, you know, if there are questions you have, if they, you can actually quantify or calculate, they would not recommend asking questions to the applicant and just wrap this up."

He referred to what I was expected to do as "donkey work."

We had data we could use to accurately reflect products now associated with a skyrocketing market. He previously encouraged me to do that, then said I could only use marketing estimates that were over four years old. So, we wouldn't use the best, most up-to-date information.

I asked, "Is the expectation that I write an assessment that supports findings of no significant impact? Is that what they mean by 'wrapping this up'?"

He sheepishly confirmed, "Yeah, that is my understanding."

He said, "Yeah. I don't know. The whole thing is so, I, it's such a fuzzy, you know, like disbelief of, Kafkaesque, the whole thing is such a Kafkaesque state of being."

Kafkaesque means "having a nightmarishly complex, bizarre, or illogical quality." Leadership sometimes, likewise, characterized what we did as so complex that it couldn't be understood. Like the magic behind the dingy green curtain, we were too plebeian to ever understand.

But this wasn't Kafkaesque. It wasn't complex or illogical, or even bizarre. It was predictable. He'd always described what we did as donkey work, sausage-making, or rubber-stamping because every CTP assessment facilitated marketing orders. That was consistent. Tactics used to maintain that were consistent. Likewise, efforts to hide problems and suppress issues that might delay marketing orders were similarly handled. That was happening again here.

He said, "I am saying that my stance, based on the same reading, is there might be impacts, but I'm not sure if it is significant," but we wouldn't look into that. In respect to potential impacts, he said, "In my opinion, it is that, what is the waste? It is packaging and it is plastic waste through use, after the use, and the sources of environmental impact are those mostly."

I asked, "What kind of plastic?"

Lacking confidence, he said, "There are two to three types of plastic."

I asked, "What kind?"

He said, "I don't know."

I asked, "Okay, alright, what is the pouch made of?"

He said, "I don't know."

I asked, "Okay, so [Branch Chief #7] and [Supervisor #7], do they know?

He said, "No."

I said, "Okay, so my secondary [reviewer], tertiary [reviewer], and supervisor are saying that I should produce an assessment, support a FONSI, and none of you know what type of plastic. And the thing is, fine, it's, maybe the plastic isn't a problem. But the fact that the expectation is to do this without actually knowing the materials, that seems really, you know, that seems a little strange to me. And fine, I'm going to do as directed, but I just want, I wanted to point out that it seems a little strange to me that, that my secondary, tertiary, and supervisor on an Environmental Science review towards granting a marketing order for [product name], don't know the plastic, don't know the materials associated with the product when we consider disposal. That seems strange, doesn't it?"

He responded, "So, you know, it's, it, yeah, we don't need to know. Again, I don't need to. I can't defend their, what their stance is. I'm telling you my stance and I can, I'm speaking for myself here, is, that yes, the plastic impacts we don't know. We don't actually know."

I said, "You actually don't know what the pouch is made of, and this, yes, so, I just wanted to, I just wanted to make sure that was clear, that you know we want to wrap this up towards a FONSI, but there obviously isn't any curiosity about what the pouch is actually made of. Just wanted to confirm that."

He responded, "I don't know that, the sources of environmental impact as such."

I asked, "Okay. What did you guys talk about?" in reference to his meeting with our Supervisor #7 and our Branch Chief #7.

He said, "This is exactly what we talked about, like, they conveying to me that this is actually at the Office of Compliance and Enforcement right now." He indicated that CTP was moving ahead to grant marketing orders.

I mentioned using information from public sources. In that case, we'd be guessing but we wouldn't need to ask the applicant. We could talk in general terms and be explicit that we weren't referencing the actual products.

He said, "You can frame it the way you want to." But he kept talking only about disposal. So, I asked if we could include considerations of use, considering the breakdown of the pouch during use, because the use of products was considered in every other assessment, and he said, "We can."

We talked again about potential exposures to microplastics during use. And we again talked about disposal when microplastics released into the bodies of users would subsequently be disposed of in their tissues and organs as well as in our wastewater systems and, therefore, contaminate ecological systems, fish, drinking water, all that stuff. We talked about applying the same approach the agency previously used in assessments to model the movement of nicotine—into a user, through a user, and through wastewater treatment systems—to microplastics. So, we had precedent.

November 8th, 2024

He then sent me an email with *his* uninformed and inappropriate determination (e.g., inappropriately citing GRAS), that I was told to reflect in *my* assessment, that there were no impacts associated with the pouch materials. We still didn't know what the pouches were made of, but I'd still be able to talk about the pouches in my assessment. He'd said, "You can frame it the way you want to." So, I responded and copied our Supervisor #7, "I'm glad we'll be able to add mention of the pouch materials." I then continued my work to do that.

November 14th, 2024

Supervisor #7 then convened a meeting with our Branch Chief #7, Branch Chief #3 and #5, and the team of scientists involved with the Microplastics Exposures from Tobacco Products Project I'd led since 2021. On September 27th, 2024, I'd provided a 37-page report to update leadership about how aspects of that project had progressed. I shared mounting evidence of exposures and about efforts to characterize effects. I shared concerns expressed at the Office of the Center Director about CTP's products flying under the radar with respect to constituents allowed in beneficial products under the purview of other FDA centers. For instance, exemptions allowing harmful constituents in a beneficial medical device or pharmaceutical might be applied to tobacco products.

That report included a summary:

Policy makers and regulatory scientists are developing draft U.S. policy to address increasing evidence of human health and environmental effects from particulate plastic polymer exposure, e.g., from microplastics, nanoplastics, and polymer-based nanomaterials. This project initially aimed to create summaries of these polymers in tobacco products alongside effects and risk information, and that is still a goal. However, early interactions with FDA and federal workgroups indicated that questionable methods limit applications of some effects and risk studies. Therefore, recent project activity has focused on tracking and communicating information to the branch, our division, and CTP about vetted science and developing policy that is relevant to CTP. These polymers are present in all tobacco product categories and exposure pathways constitute direct consumption of plastic polymers (e.g., via ingestion from oral pouched products) and indirect exposure (e.g., via contamination of food and water following tobacco product disposal). Therefore, CTP may need to address, in multiple scientific disciplines, potential problems associated with some plastic polymers as part of its compliance with the National Environmental Policy Act, the Family Smoking Prevention and Tobacco Control Act, and the Federal Food, Drug, and Cosmetic Act.

On August 1st, 2024, Supervisor #6 had referenced Office of Science priorities, "So, I don't know if you guys saw the research priorities, but they specifically mentioned microplastics." She and my Secondary Reviewer #3 had then talked about doing assessments for those 14 oral pouch products in the context of what was currently known, to reflect the best, most up-to-date information about microplastics.

Supervisor #7, our Branch Chief #7, our Branch Chief #3 and #5, and other leadership who saw that 37-page report wouldn't engage about it or the information communicated. Instead, Supervisor #7 started that November 14th, 2024, meeting with, "All right, um, so, thank you for, here we go, thank you for coming to this meeting. Um, I'm hoping that, you know, my goal is to help facilitate this project coming to a, uh it's been going on and I'm hoping for this to help facilitate an end to this project."

That was the first I'd heard of that.

April 24th, 2023

In hindsight, a meeting with my Supervisor #2, our Branch Chief #3 and #5, and our Division Associate Director provides some insight. We had that April 24th, 2023, meeting to discuss my participation on an FDA Nanomaterials Task Force that was part of that Microplastics Exposures from Tobacco Products Project. Then, I hadn't known the extent of efforts to prevent information from moving into our assessments or to cover up past indiscretions. Sharing the information I shared was part of my job and I believed that was wanted.

During that meeting, our Division Associate Director was cooperative, articulate, and pleasant. He shared his experience with nanomaterials in his previous role at FDA's Center for Drug Evaluation and Research (CDER). He talked about his familiarity with CTP's previous Nanomaterials Task Force representative, who'd left about four years prior. He said, "I'm interested in hearing about that and learning about it. I don't know a lot about it. I've heard about microplastics." He asked if I could run through the background and need.

I reiterated that nanomaterials are used in beneficial applications, like drug delivery, and that attributes allowing those technologies access in biological systems come with potential concerns aligned with the microplastic and nanoplastic pollution initiatives I'd been involved with. I talked about concerns with small particles and with degradation products of plastic polymers. I said, "When a lot of these materials breakdown during use or after disposal they can be problems for human health and the environment, and that is becoming more substantiated."

They're little, so they can sometimes move in and out of cells, for better or for worse. Those cells have things they need to do that can get mucked up when unwanted particles get in the way. Imagine trying to eat a sandwich with a mouthful of marbles.

I mentioned how I'd kept track of national and international legislative developments, from the State Department and other groups providing information via the groups I sat on. I said we were talking about "not only just the primary materials, but also any secondary substances that might be associated with those. Those materials can serve as carriers for all sorts of other chemicals, for pathogens." I said those topics are "getting a lot more attention." I said, with respect to nanomaterials, "This would be a task force within just FDA to consider the same sorts of things."

"What does the science point to?" I said, "There's been some concern that the methods that have been available aren't as informative as they could be, about the effects and the movement of these materials in biological systems. So, [considering] environmental and human health [exposures], that's improving quite a bit. And so, this group as well, is focused on, again, what legislation might be needed, how that is developing, and how do the needs of the different centers, in this case within FDA, how can they be coordinated."

He asked, "Are they covering the plastic containers that we deal with, or specific, you know, products from ENDS, or what kind of? I'm trying to picture the information and how [it] would interplay into the assessments that we do for the branch?"

I said, "The thing that really has struck me, that I've spoken with [my Supervisor #2] about, in particular, is pouch products. There are a variety of different pouch products that are becoming more popular. They're taking up greater market share, and these products are made of materials that degrade into microplastics, a variety of them. I haven't found a pouch yet that isn't. And this is something that we don't really look at. Now, do we have to look at it? I guess is one question from a regulatory perspective, and I think so. From the perspective of any sort of regulatory obligation that comes from elsewhere. So, CEQ, the Council on Environmental Quality with the White House, is talking about these very concerns. And they've really focused so far on cigarette butt pollution, and the cellulose acetate and other polymers associated with that. But there are other materials," other plastic polymers—some derived from plants—that degrade into microplastic pollution.

I continued, "To your question about different tobacco products, I don't think there's a tobacco product [category or class] that doesn't have these polymers included."

And I continued, "Back to the pouch products, and I think, in particular, this is one reason why I spoke with [my Supervisor #2] just a few days ago about potentially talking about creating a memo to the file, because these pouch products are going into the oral cavity and they are masticated. There's heat, there's enzymes, there's temperatures, and these, the more and more we learn, that those factors influence breakdown of these polymers into microplastics that are then consumed. And there's more information showing that exposure, oral consumption, for human health is a concern." I said, "In the environment, as well, these products are discarded, and so similar sorts of influences would cause these degradation products to be released." I said, "I think it's something we really need to have on the radar. And I think we need to have it on the radar not for just [our] Environmental Science [Branch], but for Toxicology, Microbiology [Branches], definitely, especially considering breakdown, microbial breakdown, because microbial breakdown is involved with breaking down these polymers. And possibly other disciplines, such as Engineering [Branches]."

Our Branch Chief #3 and #5 then chimed in, "So, monitoring the information right now is the most we could do, so we just monitor, collect information, and then we can see where to go with that. Because if we're just going to be focusing on those pouches for smokeless and for those dissolvables by itself, everything in our life has microplastics, or produces microplastics in general, and we do consume it from everything around us so maybe just compile information first and watch and monitor and then maybe we can come up a strategy for that. Because right now with very little information scattered here and there it would be very hard to come up with a strategy that will be Office of Science or CTP-wide for us for our work not just in terms of Environmental Science in terms of all other aspects. That's my opinion in general."

What?

Sure, we didn't need to jump the gun, but we needed to follow what science, the law, other FDA centers, my Supervisor #2, our Division Associate Director, and countless others already recognized as important. Our role was to inform decision makers and the public. We could, but we didn't need to do research or monitor or even compile information. We just needed to communicate what was known, and that wouldn't be difficult. With that we would include any data gaps and uncertainties. I'd already shared with her, for over a year, that we knew more than she was letting on.

"Scattered here and there"?

No.

"Everything in our life"?

Really?

Here was another weird rationale the U.S. government agency tasked with protecting the public health—with respect to tobacco products—used to exclude considerations of exposures. Her rationale precluded any concern, or even any consideration of any possible concern, by assuming an exposure associated with a tobacco product also came from elsewhere, she said because "everything in our life has microplastics, or produces microplastics in general." Not true, and her rationale meant we'd just keep piling on more contaminants because they were already present. That's akin to acting like lying is okay because everyone's doing it, when not everyone was.

And it wouldn't be hard to come up with a strategy.

"I think the science is advancing more than we know," I replied. "Considering something like the pouch products in particular, yeah, I think there might be some routes [of exposure] that we might need to pay a little bit more attention to, that might be considered important."

She asked about a memo she thought CTP produced about GRAS, constituents generally recognized as safe in food.

I said some claim "[oral pouch materials] are *food safe*," and a designation that something is food safe, doesn't mean it's safe to eat. Paper that touches your pizza might be food safe, but it might not be safe to eat that paper.

Our Division Associate Director said that GRAS memo hadn't been completed and he confirmed, "Our position is it's not GRAS," indicating that a GRAS designation for something in a food product couldn't be used to designate that as GRAS in a tobacco product. He said, "It's like saying, 'Okay, it's approved for a drug, so it must be okay to throw in tobacco.' It has nothing to do with it. It's a regular regulatory thing. But we do say it's okay to use the underlying data from some in this case. GRAS has to be publicly accessible information. That's just the way that works."

He said, "GRAS is probably one of the most misunderstood concepts, even by a lot of FDA people." He said, "You can't say 'GRAS' if it's good enough for food, it must be good enough for tobacco." He said, "Even if it's GRAS in your yogurt, it doesn't mean it's GRAS in your pizza."

Preventing staff from seeing the administrative file would make it easier to support any current thinking that countered policies like this. And completing that memo, that would put part of our process in writing in the administrative file, would make it more difficult to talk away, via inappropriate applications of GRAS constituents, potential exposures that might need to be considered.

I talked about some smokeless oral pouches we only knew were made, for example, of "four different polymers." CTP had already granted marketing orders for those products. I said collaborations with other FDA centers via the Microplastics Exposures from Tobacco Products Project helped inform considerations when we *did* know, that "this is why interaction with the FDA group is important." But even if a GRAS designation might apply or publicly accessible GRAS data was available, that wouldn't matter when we didn't know what those polymers were.

My Supervisor #2, who'd been a long-time member of our branch, recapped my experience and emphasized, "Things are starting to accelerate with, I think the top two products that are of most concern right now are these cellulose acetate-type pouches and cigarette butts, of course. I see other products, probably things like tipped cigars, maybe filt—, well, depending on the type of filter on a cigar. ENDS products, that's going to be more of a degradation issue, I think, and that's probably a little bit further down the road. But anything that FDA is going to do on an agency-wide basis is going to affect anything that we do, or how we look at it at least."

In this context, he said that scientists implementing NEPA in another FDA center hadn't been able to reach findings of no significant impact owing to concerns about pollution from a product it regulated. He said, "It did affect CDER in an environmental way where they had these formulations for sunscreens with nano[materials], titanium dioxide, and then they had to end up writing an environmental impact statement because of that."

CTP had never written an environmental impact statement because our assessments always determined that wasn't needed. An environmental impact statement would delay marketing orders and involve transparency and public scrutiny. Supervisor #2 had said we couldn't put our process in writing because it would be discoverable via Freedom of Information Act requests. How would CTP explain the process that in every other instance found no impacts, in particular when we hadn't considered what oral pouch products already on the market were made of.

In response to my Supervisor #2, our Division Associate Director said, "Yeah, I think it, I can hear you and see that this is an emerging topic that needs to be monitored." He then talked about potential guidance, about how we might need to address such exposures.

At the end of the meeting he said, "Christy, you're bringing in something that totally, it's going to have a, some complexities to it, to look at it from different centers. So, I think it's valuable. Thank you for bringing this forward. I don't have further questions. I'll follow up with you with what I hear." He said he'd share what we'd discussed with our Division Director, that he'd decide what should be shared with Office of Science leadership, who would decide how our CTP Director #2 might be informed. Supervisor #2 subsequently let me know that my participation on the task force had been approved.

In 2021, I'd begun an analysis of CTP's work with products used in the oral cavity. I summarized what we knew about the materials that made up those products and how we'd considered those materials. Smokeless products and the 14 oral pouch products I reviewed in 2024 were both derived from tobacco, but those 14 oral pouch products were assigned to a category of *other* products. However, those pouches were likely made of the same or similar materials.

I found no evidence that CTP had considered the materials constituting the pouch that users of smokeless products put in their mouths. When pouch materials hadn't been provided, there was no evidence that CTP asked for that information. To subsequently obtain that information for that 2021 analysis, we'd talked about asking manufacturers to voluntarily share those materials via a *Federal Register* request. I, instead, studied patents and other public information. Branch Chief #1 discouraged the *Federal Register* request, she said, because we couldn't validate the information provided. However, we used marketing estimates and all sorts of other information they provided that we didn't validate.

If we were to exclude considerations of product materials from regulatory science reviews, that should've been communicated to all staff associated with reviews of those products. The rationale for any exclusion should've been captured, for example, in a memo in the administrative file that, instead of being forthright about, our Director of the Division of Regulatory Project Management said required a review over 26 million documents. And all other leadership avoided the issue. But if CTP was doing things aboveboard, any such documentation should've been apparent via the years I worked on that project and via my involvement with the 2024 review of those 14 oral pouch products. I saw no evidence of a precedent to exclude considerations of pouch materials for smokeless products or for oral pouches in the *other* category.

Manufacturers were developing more oral pouch products that flooded the market without marketing orders and science increasingly raised concerns about plastic polymers likely used in those products that CTP, for some reason, hadn't considered. And if it hadn't previously, would CTP now do a scientifically sound evaluation of pouch materials in those 14 products?

November 14th, 2024

On the same day that Supervisor #7 said she'd "facilitate an end" to the Microplastics Exposures from Tobacco Products Project, my Secondary Reviewer #3 provided additional guidance about by review of those 14 oral pouch products.

He said our Supervisor #7 and Branch Chief #7 were pressuring him to complete the review, "to wrap this up." He said our Supervisor #7 called him at night, unorthodoxly outside of work hours, to pressure him. He said about a meeting he'd planned with other reviewers, to ensure our work was accurate and consistent, "So, I don't see a problem in setting up a meeting next week, but my concern is with people breathing down on my neck trying to wrap this up," and we never had that meeting.

And in respect to doing credible work, he asked, "Is it worth doing all this sort of hard process?"

Instead, we went through the draft assessment while he pointed out what should be added and deleted. He directed me to remove mention of the pouch in some instances and just refer to packaging. For example, he pointed to where I'd added in a table in the public-facing part of the assessment that the product constituted a pouch. That was a normal thing to do. It was normal to describe the product.

He explained, "This [section] is specifically for product *packaging*, it's not for the pouch per se."

He explained, "Why I did not want to accept your addition was that part is talking about packaging not about the pouch itself."

He explained, "The whole description is for packaging and not for the pouch as such."

He said, "We don't need to have 'the pouches are comprised of' there at all."

Even though we didn't know what the pouch was made of, we could've included something like, "Nicotine and other constituents delivered to the user are contained within an oral pouch."

He explained, in reference to that table, that I shouldn't reference the pouches because, "The identification is standard, that's what they want." But that wasn't standard, and he seemed to be overexplaining, as if he was trying to convince himself of something. It was standard to accurately describe the product. For example, in the same table in the public-facing part of CTP's June

14th, 2024, assessments for ENDS products under "Product Package" it says, "Silicone mouthpiece plug, silicone bottom (silicone air inlet plug), a polyvinyl chloride tray sealed on the back by foil, and a printed paper insert." The oral pouch constituted packaging to a greater extent than that mouthpiece.

It was standard to describe the product.

In the context of calculations about things like packaging, I mentioned more recent marketing data that he'd previously raised to accurately reflect the market. He ignored what I said and shifted the discussion. Using older data meant all our calculations underestimated every associated impact. He intentionally selected information that we knew would underestimate impacts.

He brought up our need to include considerations of environmental justice, that we needed to mention groups possibly more affected than others, like kids. He said, "From what I gathered, the highest use prevalence for the population right now is middle-aged and older men. White men are the predominant users of these products." He told me to add that while lamenting what we excluded. He said, "That doesn't count the kids that eat these pouches." We'd only add language about middle-aged and older white men.

He was massaging the document to facilitate marketing orders, as he described, "making the sausage" while he used phrases like "you, please, should, need to."

To indicate the applicant was in compliance with various regulations, he talked about adding environmental permit information. I pointed out that a permit had been terminated, but we didn't know why. He then backpedaled, "I'll take it back. I'm sorry." So, we wouldn't include that information. Normally we'd check, so I did and learned the permit termination wasn't associated with a violation. So, he again wanted to add that information.

He'd put the kibosh on any considerations of the pouch materials, but there were other constituents we'd talked about considering. Now that was ignored. Instead, he said we should estimate the amount of nicotine that went "down the drain" following use. He said, "I believe that we should do that calculation. I have a spreadsheet I've used which I can share with you," that we could easily adapt for these products. We talked about information that would be included in the public-facing part of the document and in a confidential appendix. We talked about how we could present that information in a way that made it easy for the public to understand. He said those calculations had been used in assessments for ENDS products, and that in this case "this product is pertinent for that" as well.

He even mentioned a threshold! He laughed because we couldn't use thresholds elsewhere. *Here* we could talk about thresholds because they wouldn't be crossed. So, thresholds could be used when they supported desired outcomes.

Estimating the amount of nicotine that moves into the environment from tobacco products is not uninteresting, and it might be important. Nicotine is addictive, and that is a concern in the context of the habitual use of harmful nicotine products. At high enough concentrations, nicotine can be lethal, but for some reason we weren't mentioning kids poisoned after eating these pouches. Fortunately, CTP had already decided not to grant marketing orders for six of the 20 products, explicitly because those flavors appealed to youth.

In CTP's assessments, considerations of constituents were selected to preclude significant impacts. Nicotine breaks down fairly quickly in the body and after it's excreted. When nicotine in urine is flushed from toilets into wastewater treatment systems, like we were considering after it goes down the drain, it is diluted and further degrades. The purpose of wastewater treatment systems is to create conditions that facilitate breakdown. Nicotine concentrations in discharged wastewater will therefore be low. The volume of water in streams and rivers where treated waste is discharged will further dilute nicotine.

In some scenarios, dilution no longer serves as a solution for pollution. In those cases, the amounts of contaminants remain too high in the environment, and therefore a threshold associated with harm might be crossed. A wastewater treatment facility failure or manufacturing facility spill could result in nicotine pulses into the environment. We could mention something like that as an informative possibility. But that nicotine, as well, would breakdown relatively quickly, especially compared with other constituents for which some proportion never will, like some of those 93 constituents we didn't consider.

Choosing considerations of nicotine outside the body, therefore, stacked the deck in favor of tobacco products. CTP diverted our finite resources to model nicotine in the environment in lieu of throwing a bone to persistent constituents associated with serious known harms.

Microplastics are also persistent and carry contaminants into our bodies and our brains. In the draft assessment for those 14 products, I'd added placeholder text about microplastics we'd update after we learned what materials made up the pouches. It was brief and objective and didn't raise concerns beyond what FDA already vetted and communicated. But that text was deleted. Someone then superimposed with nicotine considerations of other constituents.

We still had no timeline. He said he'd reached out to contacts who'd tell him when we had a timeline to review, discuss, and finalize our work products. In the meantime, I'd work on edits I was directed to make.

I said I'd raised concerns about all these inappropriate directives. And by that time, leadership from our division to the Office of the FDA Commissioner had been alerted by many CTP scientists.

November 20th, 2024

The purpose of our next meeting had been to discuss elements of the draft assessment we'd been working on. Instead, he pressured me again and again to concur on that assessment and to accept all the changes in the document. He resisted even talking about what we were there to talk about.

He'd started out offering, "Let me know what I can help you with with this."

I responded, "Yeah, so what I've done," and shared my screen. I began walking him through the changes I'd made and said, "I've been working on just making sure the numbers are right." I talked about adding needed information we'd previously discussed and the pending meeting with other primary and secondary reviewers to ensure our work was accurate and consistent.

He was agreeable to that point and said, "Yeah, no, so I think that can wait, at any point, once we have a draft that is close to done, we can do that."

His tone then changed and he said, "Couple of other things that we've talked about, and I want you to have time to address, so there are two to three broader points that I think it's worth going over, Christy, which is that I feel like what branch management wants is, is something where it is concurred and sent, to where you concurred." He was recapping a meeting he'd just had with our Supervisor #7 and Branch Chief #7. He said, "So, and that is what sort of the expectation is here." He said, "That is one, okay?"

He then said, "The second part is that we talked about a couple of other things, which is a little bit of background around the current use prevalence of pouch products in the use environmental justice portion. And then the down the drain, either using the applicant-provided down the drain data as an appendix and then pointing to that in the discussion, or doing the down the drain calculation by ourselves. So, those two factors."

When I referred to our conversations about nicotine, he snapped at me, "You brought that up as a factor for a concern for nicotine that would go down the drain!" No, he had. I quickly recapped the history of our discussions about nicotine, the fate and transport of nicotine in the environment, and concerns about diverting our attention to nicotine in the environment in lieu of more informative efforts. He then said we could exclude considerations of nicotine. He said, "That is fine."

He said, "But, you know, my point here is just that, whatever you feel is, like this product is complete, you know, and we could concur, and you concur with it."

"We're not there," I said that "we've been asked to provide a draft document." To make sure this work product was complete, I'd need the pouch materials, and ideally, not to load the assessment up with bullshit. Back and forth these discussions went. I needed to *feel* the assessment was complete, but the assessment needed to cherry-pick to support certain findings. Those two things couldn't exist in the same space.

So, I said, "There are a few things [our Supervisor #7] knows and [our Branch Chief #7] knows, that there are going to be a few things that aren't done." I couldn't concur or accept all the changes, but I said, "If you're happy moving this forward with [our Branch Chief #7], I am as well, knowing that there are a few things that need to get done."

"Yeah, let's do that," he said, "because I don't know, I have no idea what they want when they say it just needs to go to [Branch Chief #7], as I see it, it is that, you know, the primary reviewer puts it in the router and concurs it for the FONSI and then I will look at it."

I again said, "I'm not going to put it in Rhapsody [the router] and concur. No, this is not a complete document."

He insisted, "Just send me the version that you feel like we could share with them." But that made no sense. We'd shared the document with them already. They had access to it and edited it and commented on it in-a shared file. That was the only version we had, so it had already been *shared.* He wanted me to do more than share it, he wanted me to *concur* in the router, where it would be documented that I *agreed* with what had been included and excluded. I wasn't dumb and I could tell they wanted me to concur, so they'd say I agreed. I didn't and each time he said I could frame things as I wanted, I took that seriously.

I said, "We need to go over this."

But he again insisted, "Christy, can we at least have a document that has changes accepted, that has accepted?!"

I said, "We need to make sure any remaining issues, we've dealt with some way." He wasted time pressuring me to concur when our meeting aimed to work on the document. So, I said, "There are a number of things that remain. I wanted to get through a few of those today, but we just didn't." I said that it was an issue that "the pouch materials weren't in here."

His tone softened, "You know if I were in the play, I would give you a little bit of breathing room to let you finish. And this is my viewpoint."

And we came up with a plan. We'd send the draft assessment to Supervisor #7 and Branch Chief #7 and indicate we still had work to do. With respect to the meeting we had planned, he again said, "We could wait for that stage to say, that after talking to them, we decided to include," what we'd add after that meeting.

He said, "I'll be honest with you. I don't know what they are trying to do, but you are putting in a good faith effort here, and it's unfair to be sort of, trying to get at you the way they are." He said, "I'm lost somewhere, from, what is the secondary [reviewer] supposed to do?" We both knew how this was playing out violated that August 27th, 2024, training where that was described.

He ended saying, "Christy, I'm sorry, again, it's, you know, I know the intent here is to work together to get this done, but I also do not like the fact that this is getting played out. It puts people in secondary [reviewer] position in a really ridiculous shitty place where they don't tell you what they want, how, what exactly. But then there seems to be something going on. And then you're left with nothing. And there is just constant checking in, like, 'What is happening? What is happening?' What do you mean, 'What is happening?'?! Just give people some breathing room!" He was talking about conversations he'd had with leadership.

December 3rd, 2024

At our next meeting, we talked about how to navigate alarming behavior by Supervisor #7 and Branch Chief #7. We'd been asked to complete the draft review and worked together on that. Once that was done, we'd do the assessment.

He started that meeting with advice again to cut corners, "I don't think that, I don't think that, you know, there's no need to sort of *prolong* any of this."

I said, "I haven't been prolonging anything."

He said, "No, no, no, prolong *their* pushiness, that's what I'm implying." But not prolonging their pushiness meant not doing a credible review and assessment.

He talked about how our Supervisor #7 added information to our work products that wasn't accurate. She'd come into our shared files and make changes, sometimes when I was actively performing my review work. I'd politely asked if no changes would be made by others until I moved the document forward for review, but she made changes anyway. She certainly could visit those documents, no problem, but she was tinkering with our work, and it was disruptive.

This was unorthodox because she wasn't even an assigned reviewer on this assignment. Each assignment had assigned primary, secondary, tertiary, and quaternary reviewers in the system so the efforts and involvement of individual regulatory scientists could be documented. Primary reviewers were supposed to work independently, then pass that work product along to the secondary reviewer for review and concurrence, then to the tertiary, and so on. If changes were needed, the document would go back to the primary reviewer, then the process would repeat. That process was important because, although it didn't provide guidance about how information was selected or considered, it was a means to document who did what and when, and it increased the odds that each reviewer could do independent work.

He talked about wild claims by leadership who criticized staff who didn't do work when they weren't at work. He said, "I find this sort of infuriating, especially when somebody is on leave, you can't judge whether they're on the document or not to see if they've done work when somebody's on leave."

We were torn about how to proceed. He'd repeatedly raised concerns about her role and how it disrupted our work. I let him know I was reporting what was happening. Our Office of Science Director #2 had written that he was having meetings about concerns reported to him via his open-door policy, and that he'd circle back. Unfortunately, he never did, but at the time we believed he'd soon realign our credibility, like he'd said from his arrival he'd do.

My Secondary Reviewer #3 said he was trying to help, saying, "I am trying minimize where you are going to engage with [our Supervisor #7], okay?" For example, when I asked about addressing a comment by our Branch Chief #7 in an appropriate way, he said, "No, don't. I don't think it's worth going there. We just have to put some shit down here."

I said, "Okay."

He apologized and said, "You know how this is going to go. They will double down." Throughout the call he seemed frantic to put some shit down that placated them. That draft review and the draft assessment we'd complete to follow felt less like my work products every day.

I was dumbfounded, and said, "I think the way [the review] is written now, I think it, I mean, I feel like it should be okay, uh well, from the extent that, that we've, we've, we've provided something within the context of what [our Branch Chief #7] looks like she's wanting. Yeah."

Why not simply take my name off and use the names of those responsible for those work products? Why? Because they were faking independent scientific reviews. Like my Secondary Reviewer #3 had said, "At many levels, they are just using our scientific integrity to get through shady documentation." But did we still have scientific integrity? We had *credentials* that they used.

And now, *he* directed me how to respond, and even how to write my emails. For example, he told me, "We could just, if you could just respond to her and say we," did this or we did that. He said, "I will tell them that they can take a look at it and this [draft review] can be waiting in the wings," with a few issues to consider, "whenever they say finalize it and submit it."

December 11th, 2024

A colleague in the Office of the Center Director forwarded to me and others in our branch an email, "[Product name of those 14 oral pouch products] and Nicotine Pouches: What Local Tobacco Officials Need to Know."

She wrote, "I attended a webinar on Monday hosted by the National Association of County and City Health Officials and the Truth Initiative. It focused on [the product name of those 14 oral pouch products] and other nicotine pouches as *National Youth Tobacco Survey* results show they're the second-most used product among middle and high schoolers. The webinar included some info on nicotine pouches as products and mentioned that [the manufacturer] is investing $600 million to build a new [product name] manufacturing facility. There were also comparisons of [product name] pouches to [product name] e-cigarettes in marketing, occupying the market, and enforcement and compliance."

I forwarded the email to my Secondary Reviewer #3 and wrote, "Let me know if you believe we should discuss this within the context of our current work." I followed and compiled evidence from those sources. He didn't respond.

December 12th, 2024

The Technical Project Lead then wrote that marketing orders would be granted for not only the 14 products I'd been directed to evaluate, but for all 20 products, including the six flavored products previously excluded because of their appeal to youth. He asked me to update our review to reflect those changes by December 17th, 2024. I added the product names and descriptive product information, and we'd need to meet to consider those six flavored products CTP previously decided shouldn't receive marketing orders.

What might've changed since CTP had decided to deny those six marketing orders? Had the science changed, or something else?

December 20th, 2024

When I went on leave on December 20th, 2024, we were awaiting news about the draft review, and about meetings, deadlines, and a timeline.

27 | How Executive Branch Employees Become Radicalized

AT THAT SAME DECEMBER 3RD, 2024, MEETING, my Secondary Reviewer #3 said he yielded to the sausage-making because, "I grew up in a different context, in a, you know, close to poverty context, and I don't, I can't think about not having a job and being able to support kids."

Why should a parent ever be in that sort of situation? These sorts of compromises should never happen, especially in the context of tobacco products and a regulatory agency with a public health mission.

And why was I expected to follow suit? To risk my job, career, and livelihood to accommodate what *he* chose to do? He had a spouse with a job. I didn't. He said he'd make the sausage because he had kids, but the sausage-making prevented us from serving our public health role, for kids. Of course, he could protect *his* kids using the inside information that he had—to heck with other people's kids.

And those tradeoffs, purportedly for the long game, didn't come without costs. Staff talked about that rationale that meant CTP's indiscretions shouldn't be raised because that would undermine the *good things* we did, like the nicotine rule that would limit the amount of nicotine in some tobacco products.

However, even some of those efforts were questionable. Reducing the amount of nicotine in a cigarette might influence addiction in users. However, CTP staff raised concerns about increases in exposures to carcinogens and other harmful constituents when, to obtain the same amount of nicotine in reduced-nicotine cigarettes, Americans would smoke more. The industry tolerated CTP's development of the nicotine rule. Smokers would smoke more cigarettes, so they'd still make plenty while they pivoted their investments to products, like ENDS and oral pouch products, where nicotine levels are steadily increasing.

Anyway, undocumented negotiations and dealmaking might've been the most erosive behaviors at CTP. Employee records included evidence of regularly completed trainings that proved we'd been provided and understood, via competency tests, required high standards. Imagine when leadership then said the good we did came via marketing orders we denied, so we didn't need to do what was required when marketing orders were granted. Imagine when they said the law we were hired to implement didn't have teeth and was only administrative, so why bother? Imagine when they said we didn't consider product use and human health to exclude required and important considerations of product use and human health.

When staff followed that undocumented, esoteric word of mouth from leadership they trusted, because they'd initially seemed welcoming and supportive and chummy, they'd quickly become complicit. Leadership then pointed to where staff previously signed and concurred to coerce continued corrupt cooperation.

Compromises where Executive Branch employees said they needed to violate policy—or break the law—to stay employed weren't uncommon.

Higher-ups said the same thing.

Fast forward to March 6th and 7th, 2025, when our Office of Science Director #2—our head of science—said he needed to break the law to keep his job. During those meetings, illegal actions that leadership agreed to carry out were discussed.

One impetus of those conversations was a March 3rd, 2025, email to all of CTP from our CTP Director #2:

CTP Colleagues,

As you may have seen, a media outlet called The Examination *is sending emails to CTP staff and approaching people on social media asking for comment on the center. Please note that our communications team is aware of these particular requests and no action or reply to the reporter is necessary on your part.*

As a reminder, per longstanding FDA policy, clearance is required before talking to the media. Accordingly, CTP staff should not be replying to reporters without formal clearance. If you receive any emails from the news media, please follow appropriate procedures by alerting CTP's press team at CTP-OHCE-STRATCOMM@fda.hhs.gov so that they can appropriately triage the request.

Thank you,

[CTP Director #2]

The blanket kibosh on all communication was believed to prevent outsiders from learning about our indiscretions. At the same time, the second Trump administration was eliminating Executive Branch employees. So, delivering an overly restrictive message like this, then, was additionally concerning.

During a March 6th, 2025, Office of Science meeting, a union steward said, "All of our contract rights were violated, all at once through verbal orders, delivered by all of the management in FDA, and we should be very—to be clear—that those were illegal actions and contract violations that were taken by every level of management."

He said, "You got an email earlier this week. We want to remind everybody that you have First Amendment rights. You have the right to discuss your labor conditions. It is a crime to interfere with your ability to discuss your labor conditions, and people forget that when something is unconstitutional that doesn't just mean it's unconstitutional. It means it's illegal and it's a crime to do those things."

Nonetheless, our Office of Science Director #2, who knew about the sausage-making, reinforced the kibosh. He said, "Again, [the CTP Director #2] sent out a reminder about staff communicating with the press, and he has said that the stories that come out about CTP have undermined his efforts to advocate and use the channels that he has. So, everybody has to make their own choices, but that is something that I think is important for people to consider about his own efforts to advocate for us through the channels that he has, because he has every interest in protecting CTP and all of our staff so that we can do the work that we're trying to do."

This wasn't about protecting staff or the American people, that ship had sailed. This was about hiding indiscretions. This was damage control—"triage."

At the same time, Supervisor #7 directed us to misrepresent our work in responses to internal inquiries, unorthodoxly offering up her personal mobile number to chat. Because I'd opted for the Deferred Resignation Program, I was exempt from lying about my work, but it was fascinating to watch. Staff were directed not to report work loosely related or unrelated to our tobacco regulatory role. For example, two colleagues co-coordinated an unrelated meeting about vector-borne disease. That was a big effort, but our work was funded by fees applicants paid when requesting marketing orders—for *tobacco products*. Our CTP Director #2 told Congress that our tobacco regulatory work was delayed by finite resources, that the public health wasn't adequately served because we didn't have enough money for enough staff. How would it look to have staff he claimed were overworked sideline tobacco regulatory work for other efforts?

Staff delivering directives to lie and abiding by directives to lie—that violated our oath—would be culpable, and therefore might find it difficult, moving forward, to raise concerns.

This was a train wreck.

Biden era CTP leadership and staff complained incessantly about incoming corrupt practices associated with the second Trump administration. I'm sure some were truly concerned about science on the chopping block. No doubt others were excited the cherry-picking, sausage-making, rubber-stamping, donkey work might be easier. But imagine how the second Trump administration could weaponize wrongdoing in the FDA under the Biden administration.

Leadership was freaking out, and some got weird. Supervisor #7 chastised the people who voted for Trump, saying, "the same people voted wrong again." She went on and on, referring to the second Trump administration as "the second coming." She talked about how horrible life at the FDA was during the first Trump administration while she did the same things she criticized them for doing. She and others compromised CTP's credibility and mission and were now claiming higher ground. I didn't vote for Trump, but I didn't want to hear this sort of crap at work, especially as an Executive Branch employee.

Biden era CTP employees had used corrupt tactics to hide corrupt practices that happened during the Biden administration. However, some of those corrupt practices had begun during the first Trump administration. So, Biden era employees had chosen to continue corrupt tactics used during the first Trump administration, then claimed higher ground when Trump returned. And some corrupt practices, like always developing assessments that supported findings of no significant impacts, likely started during the Obama administration. Now, Biden era employees were preventing the incoming second Trump administration from discovering some corrupt tactics implemented during the previous Trump administration, because they'd continued them.

Wowie!

Who's to blame?

Everyone.

Administration after administration failed to correct previous indiscretions, and instead implemented new indiscretions to hide past ones, and inadequate Congressional oversight allowed it to continue. The union warned that the FDA would shift tactics immediately following the 2024 Presidential election, knowing the hearing of any indiscretions could be delayed until after Inauguration Day, and therefore second Trump administration tactics would apply starting in early November of 2024.

And the lies ramped up after November 5th, 2024, lies in writing that could be disproven and that were, but that were repeated, nonetheless. The subpar science we used suffered further. They behaved as if there was no recourse, and there was no evidence that there was. For example, our Office of Science Director #2 saw written evidence of lies and shoddy science via the open-door policy he repeatedly advertised, but those lies and shoddy science were allowed to continue.

On March 7th, 2025, we had another Office of Science meeting. It was a Q&A, and a colleague wrote in to ask why leadership was choosing to break the law. Our Office of Science Director #2 said, "Well, there is another one [a question] that came in about, you know, suggesting that leaders, I guess such as myself, are obeying illegal orders and that maybe we should be pushing back against those. I guess it's a question of, do you still want me to be the Director of the Office of Science? Because when you think about what I might do or not do in terms of following directives, it could also put my job in jeopardy too. So, I think there's a balance here with situations like this. So, I think that's true for CTP leadership, in general. So, kind of a downer note to end on, but I think—"

Holy shit. He'd decided "obeying illegal orders," otherwise known as *breaking the law*, was something he'd do to keep *his* job. And how far did that behavior extend? To heck with public health and the kids. For our Office of Science Director #2 to feel he should violate the law to keep his job was one thing. Telling the entire Office of Science, hundreds of employees that he directed and where our comprehensive and robust regulatory decisions were supposed to happen, was another. Our Office of Science Director #2 indicating he would violate the law to keep his job, while Executive Branch employees were losing theirs, was additionally concerning. How did that not suggest that all staff should obey illegal orders, otherwise known as breaking the law, to keep their jobs?

He was abruptly interrupted by another member of the Office of Science Leadership Team. She quickly deflected, "So, as a reminder you really want to sort of think about what to do like sort of prioritize the balance of the sort of information that could cause you stress, and you know balancing that with things that give you lightness and sort of ground you, okay? And that's different for every single individual, but you know it's easy to go down like a spiral, and you know, you want to take care of yourself, your whole self, and so keep that in mind."

What? No worries, not breaking the law is what has kept me from going down a spiral. Maybe they should try that and advise staff to obey the law instead.

She continued, "Okay, with that, thank you so much."

Our Office of Science Director #2 then said, "Thanks everyone," and the meeting ended.

28 | Hindsight

I WONDERED WHY I'D BEEN ASSIGNED as a primary reviewer of a group of products, considering my reticence to be complicit, that CTP wanted rubber-stamped. I suspect timing might've had something to do with it.

By mid-2024, I'd been assigned just one new review for the year. That was odd. Review assignments were tracked on a spreadsheet to ensure work was equally distributed, in part because leadership in our division previously denied work to staff that raised concerns. So, maybe it was difficult to deny work for certain staff.

CTP also wanted primary reviewers of PMTAs to have had previous PMTA review experience. Those 20 products were submitted for marketing orders via the PMTA pathway, and I had PMTA review experience. Team 1 was on the rotation to receive an assignment and other Team 1 primary reviewers, who also had that experience, might not have had room on their plates.

It was also possible that any concern raised by any primary reviewer simply didn't matter. They'd form work products into what they wanted and tack our names on. Perhaps leadership knew their efforts to form our work products to always support marketing orders—the sausage-making—would always be supported.

May 22nd, 2024

If she had anything to do with it, I also believed my Supervisor #6 had wanted to do work aboveboard. She knew it violated our policies to implement and hide CTP's fuzzy math, and she identified other problems early during her time with us. She talked about other evidence in our division where input that should be considered from scientists was not, and for that reason she said staff left CTP. With respect to what was happening in our branch, she said, "I'm not exactly sure what's behind that, right? It's part of the reason I kind of wanted to, you know, look behind the curtain, so to speak."

She was concerned that our process always reached findings of no significant impact, that the process that always led to the same findings hadn't been described and said, "There's no guidance for our branch." She said about procedural documents, "I wonder if it's possible that we don't have any." She said, "I wish there was some way I could come in and be like, 'Oh, here's the documents,' or know where to find them or if they exist. But I do have the feeling that maybe it's something that hasn't been done yet, so—ah, man."

We talked about common references to litigation and our Division Director's expectation that we compromise our work because, he said, "We can only do one thing at a time. We just need to meet our review deadlines."

I indicated that my role protecting the agency from litigation was by doing credible work, not by cutting corners and repeating and obfuscating previous indiscretions.

Supervisor #2 had said back on December 14th, 2023, that we couldn't put the specific process we used to develop our assessments in writing because it would be discoverable via Freedom of Information Act requests. That was concerning on its own. At that time, it didn't strike me that might've applied more broadly across our regulatory work. Other clues increasingly revealed it might, like CTP's use of industry-reported data, studies, and conclusions, and sidelined efforts to produce memos about things like GRAS, air emissions, and lead exposures that would've provided credible benchmarks we'd need to appropriately consider.

CTP was to develop and include procedural documents in the administrative file. Despite the requirement via 21 *CFR* § 10.70 that employees be provided access to the administrative file associated with their work, CTP wouldn't allow that and played weird games to obfuscate those requests. Maybe those documents didn't exist. Maybe leadership didn't want us to see that what was outlined there differed from what we did on the ground.

Who knows?

FDA would need to demonstrate how each of its centers was in compliance with standards meant to ensure we did credible work. CTP would report that to the FDA. The FDA would pass that along to HHS. HHS would pass that along where it needed to be passed along.

Was Congress paying attention?

Ubiquitous external messaging suggested leadership might've been feeding inaccurate accounts of our work into the administrative file and via other means of reporting. For instance, that 2015 report "National Environmental Policy Act; Environmental Assessments for Tobacco Products; Categorical Exclusions" talked about our process. It described our process "to ascertain the environmental consequences of [our agency's actions] on the quality of the human environment and to ensure that the interested and affected public is appropriately informed" as "an integral part of its decision-making process."

Integral is defined as "essential to completeness." However, it wasn't integral. It came after the fact and therefore had no bearing on those actions it was meant to objectively ascertain the consequences of. For instance, I'd been told CTP had decided to grant marketing orders for those 14 oral pouch products and, therefore, that I'd need to develop an assessment that said there were no impacts associated with those 14 products. That wasn't an "integral" part of the agency's decision-making process. The decision had already been made. Then, all of a sudden, I'd need to add six more products, because CTP now decided to grant marketing orders for those products too.

Did anyone bother to confirm what CTP said happened really did?

CTP and FDA knew it didn't.

And the White House knew. At a meeting of the National Association of Environmental Professionals in Minneapolis, Minnesota, from May 5th to May 8th, 2024, an individual who'd worked at CEQ, the White House Council on Environmental Quality, approached me and my colleagues to ask how things were going. He said he talked frequently with our Branch Chief #1 and offered condolences about problems we needed to navigate at CTP.

Did Congress know?

Was Congress paying attention?

May 29th, 2024

Supervisor #6 started our next meeting asking, "There was a toxicologist who filed a whistleblower complaint against CTP, I believe it was 2019. I'm not sure if you're familiar with that?"

A complaint from our division had been raised to the U.S. Office of Special Counsel, the federal agency tasked with ensuring good governance within the federal government. The complaint described wrongdoing and retaliation, in part, associated with a 2019 memorandum that directed scientific reviewers to dumb down their determinations. Findings indicated that regulatory science determinations in our division involved a variety of unscientific practices that included "eyeballing" data.

Yikes!

And leadership had retaliated against staff who raised those concerns.

On December 15th, 2022, the U.S. Office of Special Counsel (OSC) alerted the President and Congress about wrongdoing in our division. That disclosure found that CTP "had relaxed its standards of review for certain tobacco products and stifled attempts by its scientists to raise concerns." The disclosure continues:

A leading CTP toxicologist-turned-whistleblower disclosed to OSC that a 2019 memorandum issued to CTP's scientists revised the process for evaluating potentially harmful ingredients in substantial equivalence applications used for certain tobacco products. According to the whistleblower, the memo directed scientists to stop using objective, quantitative data to evaluate applications and to instead use an approach which was more akin to "eyeballing it," resulting in unclear review standards and less reliable decisions. The whistleblower further alleged that flaws in CTP's internal scientific dispute resolution process effectively prevented the whistleblower and several other concerned scientists from raising these issues within the agency. This internal dispute process is intended to safeguard the integrity of CTP's scientific evaluations, including consideration of certain tobacco products to ensure those authorized for market meet the applicable health and safety standards.

OSC referred the whistleblower's allegations for investigation. In response, the FDA convened an independent panel of scientific experts to evaluate allegations concerning the allegedly flawed tobacco application review process. The panel largely agreed with the whistleblower's concerns, finding that the process lacked "quantifiable standards or criteria." The expert panel made six recommendations, including a new process for resolving scientific discrepancies, which the agency largely adopted.

In response, the whistleblower acknowledged CTP's progress in giving scientists discretion to consider quantitative data in reviewing potentially harmful chemical compounds in substantial equivalence applications for new tobacco products. CTP also revamped its scientific dispute procedures and created a mandatory training program for all CTP staff involved in scientific decision making. While supportive of these steps, the whistleblower stated that FDA scientists will still need courage to challenge a system that "discourages dissenting voices."

"The public depends on the FDA to vigorously implement and enforce our nation's health and safety laws, especially when new tobacco products are being brought to market," said Special Counsel Henry J. Kerner "I am deeply troubled that FDA's own scientific dispute process failed our whistleblower and fellow concerned scientists, and as a result failed the public. I commend the whistleblower for persevering in the face of these unnecessary obstacles. I also thank the FDA for taking the allegations seriously by convening an independent expert panel, and for taking positive steps to repair its review process."

Unfortunately, corrective measures failed to address the extent of CTP's "eyeballing" and other wrongdoing. The new Scientific Dispute Resolution (SDR) process that our CTP Director #2 manipulated to quash concerns about fuzzy math and inappropriate leadership behavior was a required outcome of that investigation. Another required outcome was our division's production of memoranda that would clarify the scientific basis behind CTP's regulatory decision making. Those were the two memos our Division Associate Director said staff didn't have time to prepare and that our Division Director referenced when he said we could do only one thing at a time. Those two memos did nothing to clarify our process—they literally had nothing to do with it—and CTP still sidelined efforts that put our real practices in writing. They'd learned their lesson after putting things in writing in that "eyeballing" memo. Leadership, instead, doubled down and continued even more of the same sorts of indiscretions when staff tried to do work aboveboard.

And where was Congressional oversight?

Efforts to address—or even identify—the breadth of CTP's problems were insufficient. Our new CTP Director #2 and new Office of Science Director #2 had recently come on board from the outside, but complicit lower-level management remained. To cover their butts, complicit staff weren't always forthright. Their supervisors might've protected them when what they'd done went unnoticed under their watch, or they'd directed them to do sketchy work.

Regardless, the same behaviors not only continued, but escalated. That first directive that I received to lie came in April of 2023, bold considering the disclosure of wrongdoing and mandatory training to ensure the integrity of our work happened just four months prior. Behavior by the same leadership involved with previous wrongdoing transformed into transparent and ridiculous efforts to hide previous indiscretions.

That December 15th, 2022, disclosure of wrongdoing was later updated:

***UPDATED** to reflect that the substantial equivalence review process at issue applies to traditional tobacco products and not necessarily to e-cigarettes.*

"Not necessarily"? Why is the U.S. Office of Special Counsel not explicit? The issue either applies to e-cigarettes, or it doesn't. And all product categories under CTP's purview have involved questionable review practices. It must be made public to which products wrongdoing has applied.

During that May 29th, 2024, meeting with my Supervisor #6, I said about those findings of wrongdoing, "That is what is happening. It is the same. I just cannot believe that the memo came out, I think—in December of 2022—we developed the Scientific Dispute Resolution process." I said, "To my understanding, our SDR process was an outcome of that." I said, "It's so recent and it's the same behavior. So, I don't know if it should be surprising or not that people would continue the same thing."

She agreed, "Right."

We also talked about emails I'd just copied her on to loop her in about current activities associated with the Microplastics Exposures from Tobacco Products Project. We talked about air quality and secondhand and thirdhand emissions. We talked about cumulative impacts where emissions that accumulate over time would cause more harmful exposures. For example, we talked about children exposed to lead that, with many other harmful constituents, would build up over time. And I said, "I think as scientists, we know if we're to really address it cumulatively, it would probably point to an environmental impact," and therefore not support findings of no significant impact.

She agreed, "Yeah."

June 6th, 2024

During a Team 1 meeting, Supervisor #6 repeatedly asked us to put problems in writing and brainstorm possible solutions. She said, "I think at least having some stuff written down, having some brainstorming sessions on this, and thinking about how to make things better would, I mean, it can't hurt. So, at the very least, we'd have some sort of roadmap to what we'd like to see." She said, "I can also take notes and send it to people as needed."

She asked one colleague about one topic, "Would you be interested in putting all of this down on paper? Just in terms of, like, suggestions and ways to improve." She said, "I think having something written down, because these are all really great ideas and I think they're really important, if anyone wants to take the lead or work together." She asked another colleague, "We were talking about how having an opportunity for training with things like CEQ and NEPA and bringing in people from the outside would be really helpful. So, I wonder if this is something you would be interested in thinking about?"

She said, "There's so much to do, right? A lot of it's related. So, I think the brainstorming is helpful, and thinking things through, and with changes coming up there's going to be some opportunities to move some things forward."

Our meeting ran long, and she made a point at the end to mention the stymied effort to update the air quality section of our assessments. She said, "The project that [our Division Associate Director] had you guys do, you were looking at something that had been done previously and rewriting it with updated literature, and then the purpose of that would be to become basically a boilerplate sort of language. Was that sort of the hope of that, or what direction was that taking?"

A colleague said, "We didn't get any follow-up on that. It seems like it kind of didn't go anywhere. But that was the idea. And we put quite a bit of effort, although it's not perfect, but it's something that is a good step in the right direction in terms of updating things. But it would be good to hear back from someone."

June 18th, 2024

Supervisor #6 started our next meeting saying, "I'm afraid I get to be the bearer of bad news. Apparently today, so I spoke to [our Branch Chief #7]. I brought up the Issue Tracker, you know, talking about what people in our team are doing. She said that she was in a group who was doing the Issue Tracker and they're almost done with it."

Our Branch Chief #7 was lying, so Supervisor #6 said, "I find it interesting that she said that they were almost done with that, and that *this* team [Team 1] was working on it," and "*they* [Branch Chief #7 and her colleagues] haven't done anything on it in a year."

In response to higher-ups' interests in efficiency and consistency, I'd worked for about a year on the Issue Tracker that would support that. That tool would quickly align common problems with vetted approaches. It would document each issue, how it had been handled, and at what level. It would add a check for false negatives. Our branch leadership questioned issues we raised but did little to nothing to look for problems that might've been missed. When those problems were discovered later, purported fear of litigation prevented corrections. The Issue Tracker would help catch those false negatives and prevent inconsistencies.

For example, say a primary reviewer raised an issue with missing marketing estimates that was resolved, following discussions with their secondary reviewer, by writing a request for that information to the applicant. In that case, details about the issue and how it had been resolved would be included in the Issue Tracker. Language used in the request to the applicant would be available for other primary reviewers to use when they encountered similar scenarios in subsequent reviews. Individuals involved with reaching that resolution would be documented for subsequent clarification, if needed.

What Branch Chief #7 referenced was the effort to compile language previously used to request information from applicants. That had been incorporated into the Issue Tracker.

Issues not raised to the applicant, that were instead resolved internally, would also be tracked. Individuals involved with reaching resolution would likewise be documented for any needed subsequent clarification. We frequently had the same discussions again and again about the same issues, and that wasted a lot of time. We often didn't know about siloed considerations of similar issues that therefore led to inconsistencies in our work. For example, we might ask one applicant for information that would delay their marketing orders, but for another applicant we'd instead figure out how to work around insufficient information. The latter applicant would therefore receive marketing orders sooner.

That also opened us up to litigation and was one reason the Issue Tracker had been supported in the first place, following discussions with our Office of Science Director #2 and CTP Director #2 about the fuzzy math incident. Our CTP Director #2 had written again and again things like "the Office of Science will be updating reviewer resource materials/guides, as appropriate, to ensure consistency," and the Issue Tracker would help get us there.

I'd met with CTP staff and primary reviewers about logistics associated with creating and using the Issue Tracker. They thought it was a great idea. All our review-related decisions would exist in one database with a dashboard and functionalities that mirrored other tools we used, so the effort to get it up and running was minimal. We'd save time because reviewers would understand the rationale used to preclude a request, preventing unnecessary meetings that delayed our work products. The tool would ensure our regulatory decisions were consistent, and management or any means of oversight could use that tool to understand the rationale behind issues excluded, and therefore previously precluded from their understanding of our work. Repeated issues that might require higher-level considerations could also be easily pinpointed when they frequently appeared in the database.

Did they not want to be consistent, or did they not want to document the fact that we weren't?

Supervisor #2 had written back on October 3rd, 2023, that "[Branch Chief #4] discussed meeting with leadership so that our branch be allowed to ask for marketing estimates for the [tobacco company name] applications even though they are a [tobacco company name] subsidiary and we were not allowed to ask in the past." I never learned why some companies were exempt from requests for information when we were allowed to, or needed to, ask others.

Other things would look strange captured in the Issue Tracker. How would it look to say, "Branch Chief #4 directed primary reviewers via deceptive arguments to pretend data combined for 32 products informs regulatory decisions about 16 individual products"? How would it look to say, "The Deputy Director for Regulatory Management concluded that what is in small text can be ignored"? Or "Science Policy Branch concluded there is no issue because no issue was raised previously, even though no one noticed the issue previously." How would it look to say someone directed primary reviewers to state that the applicant stated things that the applicant didn't state? Or, that we didn't need to know the materials constituting a pouch that users suck on because "everything in our life has microplastics, or produces microplastics in general, and we do consume it from everything around us." How would it look to say, "Lead exposure via secondhand and thirdhand emissions was excluded because Branch Chief #3 and #5 said we don't consider human health or the use of the tobacco products that we review"?

New leadership's efforts to improve efficiency and consistency, that would therefore create transparency and accountability, would need to get walked back by more seasoned staff that better understood the current thinking.

One thing that was consistent were efforts to avoid articulating what we did. Supervisor #6 had done a normal thing that a new supervisor would do. She said during that June 18th, 2024, meeting that she'd asked management for information about our process. She said she tried several times with several people. She said our branch management wouldn't point to anything in writing, but said, "Oh, it's in the training."

She said, "But nobody would give me something in writing. And I was like, just, you know, write it down for me, write down your process and I'll explain it to my team. And then everyone can be on the same page and—'No.'"

She said, "I was like, okay, can you show me in the training where this is? And they're like, 'It's there. It's there.' It's not there."

She said, "Even the most simple thing, like the process for submitting a review, I can't even get that."

Instead of putting anything in writing, she said, "Other teams know all this, like, *esoteric* information that only they know, but not everybody has that same information." Those individuals, likewise, wouldn't put in writing what they did to always ensure marketing orders were facilitated.

As a new supervisor, she was shadowing my Secondary Reviewer #2 on that waterpipe review that had gotten weird because the product was also labeled as pipe tobacco. He was on Team 2 and she said they weren't letting her shadow him. She said, "It doesn't make any sense to me at all. I was supposed to be shadowing him, and I wasn't privy to anything. He didn't share anything with me."

I replied, "It's just become more secretive."

"Yeah," she said.

She came to the same conclusion many of us had. She said, "I feel like [it is] the perfect timing to catch everything up, to get everything right. Even if they've done things wrong in the past or they didn't understand something, why not fix it now?"

I agreed, "It's the perfect opportunity to say, 'Hey, new branch chief, new office director. We can even say sort of new CTP director, new CEQ guidance. All these reasons. And when I look at some of the NEPA cases, in other cases, there are cases where it's like, 'Oh, the agency was in the process of making corrections.' That could be defensible, I think."

She agreed, "Exactly."

June 26th, 2024

I was working on that review of products labeled as both waterpipe and pipe tobacco. Supervisor #6 said, "Where you talk about the labeling on it, waterpipe and stuff is important because nobody else is bringing it up, and it does make a difference." She'd come to us from a Toxicology Branch in our same division and also knew emissions would differ.

We were having difficulty with messaging about that assignment from my Secondary Reviewer #2 who was from Team 2. For example, he said we didn't use some information from the applicant because we assumed they make it up, but we chose to believe other information they'd provided that we didn't confirm. He said we stopped considering cumulative impacts in our assessments during the last administration and that he wasn't sure we'd bring that back. But it had already been three and a half years since the first Trump administration's Executive Order to take that out had been reversed by the Biden administration. We could've simply reverted to what had been done previously. Although we certainly could improve upon what had been done before, not much work was needed to at least minimally meet compliance.

I said, "There is an enormous amount of pressure to just create a crappy document here. Clean it up, clean it up, clean it up and get it to the tertiary reviewer. Clean it up. This is a serious problem as far as scientific integrity goes. I've asked [my Secondary Reviewer #2] too, in the email that I sent to him, 'What is my role here?' Is my role here just to do what the secondary reviewer says?"

Supervisor #6 replied, "I don't think so." So, her response aligned with what our Division Associate Director, Deputy Director for Regulatory Management, and Office of Science Director #2 had said about the assessment being my own. She said, "It feels like maybe this is the issue with people not having guidelines," where some were making up their own rules. But even so, why was there so much pressure from these secondary reviewers to consistently do things that deviated so strongly from our training and higher-level messaging?

We talked about recent testimony by our CTP Director #2 to Congress where he repeatedly indicated our work was delayed because science takes time, saying we did robust reviews, that as a scientist he was acutely aware of how important that was. But he knew we didn't. What kind of precedent was he setting among CTP staff, that we could lie to Congress? I said, "Our colleagues need to understand how serious our work is. When we lie in the course of our job, that's serious."

She replied, "Yeah."

She advised me how to proceed in my communications with my secondary and tertiary reviewers and said she didn't think I should sign off on a document—to sign my name to someone else's opinion.

June 28th, 2024

I didn't feel I could sit on what was happening, so I emailed my Supervisor #6 and our Branch Chief #7, "I'd like to talk with you two about next steps regarding raising concerns with our approaches. My intent is to work within the system as much as possible and issues have been known to leadership for a very long time."

I wrote, "Right now, my feeling is that problems should be shared with the Senate Judiciary Committee as they fall in line with the June 12th hearing and CTP leadership has not implemented changes despite messaging for some time. Questionable practices have not been corrected when some mechanisms have been put in place to do so."

I wrote, "I'm happy to learn about ongoing activities to improve our processes and credibility, but we are not doing what [our CTP Director #2] indicated to the committee we are. Again, I'm happy to work together and hope that we (collectively at CTP) can."

July 1st, 2024

This was a big deal and there was no response, and I got scared. So, I retracted. I wrote to them, "I believe it makes sense to wait a bit for changes to be evident, so I'll wait before deciding if/how concerns might be further communicated. Thank you both for your responsiveness and support!"

My Supervisor #6 mentioned that CTP considered labeling issues in the context of every review, asking primary reviewers what I'd always heard, too, during each review, "Is there any labeling issue?" She was referring to the products labeled as waterpipe and pipe tobacco and said, "Mislabeling a product is a serious thing."

I updated her, via a fairly long and detailed discussion, about various recent and ongoing aspects of the Microplastics Exposures from Tobacco Products Project, and I said, "I'll let you know if anything else comes up."

She said, "I find the microplastic things pretty fascinating, particularly, I mean, obviously it's something we're learning a lot about, but every day, every time something comes up, I'm like, oh wow, yeah, that's a pretty interesting project."

"Yeah," I said, "and I have a meeting on Monday with everybody [on the project team]." I said, "I've been going through the pouch product polymers. That's what I'm focusing on. And today I was going through these patents. It is so interesting to see how the industry describes the materials. They have these long lists of natural materials that they might use, and then they talk about the plastic polymers," and plastic polymers are sometimes derived from natural materials, like plants. But plastic derived from natural materials is no less plastic.

I mentioned how I'd been working with the Office of the Center Director to share with them information about plastic polymers in tobacco products. I told her we hoped to identify legislative needs we might need to anticipate and talked with her about comments I'd provided on specific pending agreements. I continued with a longer discussion about particular materials. I made the point that we couldn't confirm, based on patents, the materials that constituted the pouch in the products we reviewed. The industry indicated in patents that the pouch could be made of a variety of materials.

I confirmed what she said, and let her know I was working with Engineering, when she pointed out, "I know that we hardly ever think about them, I mean polymers. I mean, I guess Engineering might do a little bit, but I don't feel like they really think about it."

She ended the call mentioning that she'd soon schedule my mid-year performance evaluation where we'd talk about how my performance met the duties and expectations of my position, as well as plans for the next six months. She said, "I'm still waiting to get, I have to get information from [my Supervisor #4] and [our Division Associate Director who was my Supervisor #3] and [my Supervisor #5]." She said, "I'm supposed to compile everything. I'm gonna try to compile everything and do the mid-year performance evaluation meetings hopefully next week."

July 22nd, 2024

In the meantime, she'd assigned me that review of those 20 oral pouch products. My Secondary Reviewer #3 had said CTP had already decided to grant marketing orders for 14 of those 20 products and needed assessments for those 14 products. He said CTP likely wouldn't let us raise any issues or ask the applicant any questions, but that I'd still need to raise any issues that needed to be raised. The outcome of the assessment couldn't be determined until the assessment had been completed, yet CTP determined the outcome beforehand.

July 30th, 2024

My Supervisor #6 had scheduled my mid-year performance evaluation for July 30th, 2024. We were expected to sign off on those evaluations by July 31st, so we didn't have much time. I arrived ready with my list of accomplishments and ideas to discuss for future work.

Instead, she started that evaluation saying, "Well, I think you, I don't know if you had a chance to talk to some of the other people about how the mid-year evaluations are going. There's basically, what's happening right now is, I think, I figured out that they, that I'm here basically to be the messenger. I kind of feel like maybe that's why they had this [temporary supervisory] detail filled, for the mid-year [evaluations]. That's my feeling on it right now. So, I think you can probably imagine how this might go."

What?

I'd imagined our meeting would constitute a proper mid-year performance evaluation. I imagined it would be like what I'd experienced during the three other mid-year performance evaluations I'd previously had at CTP.

"Oh, I haven't," I replied, "I've heard a little bit, but not really. So, I don't really know."

She talked for a long time before sharing what she said she was expected to share with me. This was difficult considering this meeting was about my employment. She kept talking about something I hadn't heard yet. So, I said, "I haven't heard what it is yet, but my understanding is that if there's anything that is a problem it needs to be in writing at the mid-year. I don't know what it is yet. I just don't know. I have done what I'm supposed to do. So, I'm curious given that it is the 30th and I haven't received any negative input. I'm curious, because we're supposed to get that ahead of time. So—"

She interrupted, "This was the same issue that another person on the team had. Basically, they had like a couple of things that were brought up and they were like, 'You know, this would have been helpful if it had been brought up at the time.'" She said other members of Team 1 had been unfairly criticized in their mid-year performance evaluations, for example, for "not following the procedure," when, she said, "we didn't know the procedure." She continued, "You know what I mean?"

I still didn't, because she hadn't told me.

She continued, "So, it's stuff like this. So, I just want to put that out there because I'm going to have to share this with you, and I've had to share with a couple of people things that, you know, I'm sharing. I'm the messenger. So, I, anyway, I'll go ahead and when you're ready, it's fairly lengthy. So, if you can bear with me, uh, are you ready?"

What?

Ready for what?

I replied, "I guess so?"

She then read a bunch of lies about my performance from a narrative she finally agreed to share on her screen, and she said she knew they were lies. She said primary reviewers were being "warned," that this served just as a "warning" because it wasn't "in the record."

But how those lies were delivered, because they weren't "in the record," made that more concerning. If she entered those lies into the record, there'd be a record of those lies. And I could also refute those lies—and documentation proving those claims would need to be provided. Instead, she said it wouldn't be entered into the record *yet*. She said she just wanted us to see what was being said about us that might surface again at our annual performance evaluation that would happen in January of 2025, so we could prepare.

She said, "I am doing this as a favor to my team because I got, I want you guys to be able to see what's being said so you can reference it at the end of the year, because I don't know what supervisor is going to be here at the end of the year. I don't know what they're going to be given, what information they're going to be given. And I think it's only fair that everyone on the team sees what's being said. And that's why I have put it in writing. So, in terms of why it's in writing, it's, it's a choice that I made. Because I feel that that you guys need to see this."

She'd agreed to be the messenger of lies about employees' performance—that violated policy. And she refused to conduct a proper evaluation. Yet, she acted like she was helping us. She lamented as if she was a victim, yet she'd agreed to be complicit.

The Performance Management Appraisal Program (PMAP), that this mid-year evaluation and the subsequent annual performance evaluation were part of, supported the agency's decision to continue or terminate one's employment. Therefore, there was a handbook and training to ensure it was done right. Corrective actions were to be provided at the time an employee did something that required improvement. Such communications were to be timely and include sufficient detail, and who made those criticisms was to be explicit. When that information hadn't been provided, we were to ask for those clarifications. How else could we address criticisms properly?

I'd not received any criticisms—verbally or in writing—during my time at the FDA. I'd consistently performed *above expectations*. A 3/5 rating indicated an employee met expectations, and I earned a 4/5 rating in 2021, 4.2/5 in 2022, and 4.2/5 in 2023. Despite conflict surrounding those questionable behaviors and practices, I'd been thanked for raising concerns as our training required and encouraged us to do. Training said we should embrace conflict because the public health relied on scientists raising concerns and hashing things out.

I'd done more, even by then, in 2024. I've reported contentious stuff here, but there were many productive interactions that were collegial. I'd previously received credit and kudos for the same and similar work that progressed in 2024, for example, for developing and coordinating the Issue Tracker, for my role in our WOMEN in CTP Core Leadership Group, for greenhouse gas compliance efforts, and for helping coordinate meetings and events. Post-event metrics indicated an Office of Science event that I coordinated, attended by our Office of Science Director #2, was a great success! I also previously received credit and kudos for my work with microplastics that progressed substantially in 2024.

In 2024, as I'd previously done, I'd also responded quickly, been proactive, submitted work products under my control before deadlines, etc. When there'd been contentious communications, I'd followed directives to engage as our Office of Science Director #2 directed. But I was patient, for example, waiting at least two weeks before circling back, considering workloads and how pressing issues were. I always offered more info. I said, "Thank you!" and offered to meet and work together to meet our mission. I regularly communicated my understanding that applying science with so much uncertainty in regulatory contexts was challenging. I'd been thanked for raising issues and did so within proper channels. None of that was mentioned in the dishonest narrative she showed on her screen.

What was happening violated policy and she knew that. It was blatant, and that pointed to much bigger concerns. My understanding had been that the aboveboard functioning of our government required Executive Branch employees carry out the functions of their employment without fear of reprisal. She showed us, instead, that management would lie in writing about our performance as part of a process that influenced our employment. She said she was the "messenger" that showed us that management could lie in writing. She was passing along a threat. Leadership felt empowered enough to create a written record, that she showed me on her screen, of easily disproven lies that threatened the livelihood of employees expected to abide by unwritten policies that always facilitated marketing orders for tobacco products.

Like others before and to follow, throughout this one hour and 14-minute call, she shifted between Dr. Jekyll and Mr. Hyde.

I said we'd need to address each false statement. The criticisms lacked context and it was unclear from whom they came. Regardless, they could be easily disproven and I started doing that. I asked, "Who wrote this?" I was expected to request that clarification.

She said, "So, this is like a compilation from the previous people that have been your supervisors. I got input from [our Division Associate Director who was also Supervisor #3], [Supervisor #4], and [Supervisor #5]." She said, "[Supervisor #5] didn't have much to say, but she wasn't there very long. So, you can probably imagine who."

Imagine?

I said, "This needs to be separated out by supervisor, first of all, if I'm going to respond to this. I'm sensing some serious retaliation here. It is the 30th. And this is the first time I'm seeing this. And I don't even know who this is from. I need to see what supervisor this is from. I need to see who it's coming from. So, let's start with that."

She then dismissively scoffed, "Well, that's going to be difficult to do."

Understandably, it was uncomfortable being lied about, in writing by unnamed leadership during a performance evaluation, as an Executive Branch U.S. federal government employee where there'd been so much strife surrounding our tobacco regulatory work. So, I said I wanted someone else present.

She replied, "Okay, that's fine," then steamed on, regardless, as if I'd said nothing, agreeing to then ignoring that request. Like others before and to follow, she was obstinate and argumentative and then would calm down and reveal what was happening behind the curtain, finally divulging, "I will say that [our Division Associate Director] compiled everything and put these together and sent them to me. So, that's who."

Why would someone at his level, who so frequently said he was so very busy, take the time to concoct and compile narratives for mid-year performance evaluations for employees three levels below him when those statements were supposed to remain parsed by supervisor? And he was a long-time FDA employee, so he knew that any criticisms needed to be specific, to include sufficient detail. Why had he chosen to be so vague and dishonest?

I said, "This is a retaliatory performance characterization. It is inaccurate. You know, I was hoping this would go differently," and said on the verge of tears, "I'm upset."

She replied, "Yeah, I wish I could give you, and this is why, and this is part of the reason that I'm, I've been sharing everything in writing with people. I feel like they're kind of throwing me under the bus here because—"

I interrupted, "*Your* performance is not being mischaracterized. Mine is. This is *my job*. I am a sole provider and I am 53. Okay? You are a supervisor. You are not being thrown under the bus. I am. I have been here for over three and a half years, and I have done my job explicitly as described. I have reached out how I should in every instance, and this is disgusting. [A colleague] is being thrown under the bus. I am being thrown under the bus." I said, "Everybody who knows about this and does not say something, and does not raise it, is complicit."

I said, "We have a procedure for a PMAP, right? And PMAP is about our employment. There are very few other things that affect our employment. Our employment is our wellbeing. This is not a country that has, you know, provisions for individuals based on their existence. So, that's why we have a procedure." I asked, "[Our Division Associate Director] provided this, but what was his intention, do you think, did he just send it to you and asked—"

She interrupted, "Yeah, yeah. So, I mean, and this is the same, honestly, this is the same thing I saw with some of the others. There's like stuff that's, just, where does it even come from?"

I dispelled a false claim in his dishonest narrative and said, "This has been written and described a number of times. This is actually unbelievable. And I think you've seen these emails" that prove that.

She agreed, "Yeah."

I talked about specific work that our [Division Associate Director] knew about that was important and that he said was important and valuable, that he thanked me for, but that he excluded. I said, "There's nothing in here about that."

She agreed, "Yeah."

We talked about other good work I'd done and I said, "I do a lot of stuff. There's none of that here. It's all criticisms."

She agreed, "Yeah."

I said, "There's nothing positive."

And she agreed, "Right."

I said, "I mean, there's nothing. I have worked incredibly hard this year at raising serious concerns."

And she agreed, "Yeah." And we talked more about the lies.

For example, he wrote "things about me being inefficient."

I said, "not true."

And she agreed, "Yeah."

She said, "I'm sorry."

I said, "Something should be in writing that I can lean on at the end of the year, and I don't need like super kudos. I just want it to be accurate."

She replied, "Yeah." But she wouldn't do it.

I said, "I deserve an accurate and comprehensive PMAP evaluation. It's the 30th. We need to get this done properly. So, if that can happen tomorrow, that would be great."

She then shapeshifted, becoming petulant and dismissive, "So, I'm not going to be holding this meeting again. I have covered the mid-year PMAP and this was—I know that people are not hearing what they want to hear and I shared it with you because I think it's important that you know that you are being set up here. But in terms of getting people to provide more than this, it's not going to happen."

Wow, nasty. If our work was aboveboard, why would they feel they needed to do all this? Wasn't it illegal to threaten Executive Branch employees with lies?

I asked, "So, you said it's important for me to know that I'm being set up. Is that what you just said?"

She warned, "They're letting you know that, at this point, it looks like, I mean, when I see words like 'inefficient' and basically like 'delaying,' what is it, 'leading to unnecessary delays' [that she knew were lies]. I feel like this is warning you that you're basically on thin ice. And I'm trying to think what I said right away when I saw them. There's, I mean in terms of trying to revise this and change it, this is basically not—I'm not in charge of the PMAPs here. They're, they're making me be the messenger. So, I could change it and give you a completely different PMAP tomorrow, but it won't change what they, what they're saying to you. You, you understand what I mean?"

She said *again*, "They're setting you up here. And that's why, I think, what I'm seeing from this is that they're saying, 'Oh, you're asking too many questions.'"

I said, "What this is doing, is basically saying, 'Shut your mouth and don't ask questions'?"

She confirmed, "Yeah," even though she'd repeatedly directed us to do that.

At the end of the call I said, "We all sign that job oath. We're not working for [our Division Associate Director]. We're not working for [our Office of Science Director #2]. We're working for the American people, and that sounds super cheesy, but it's true. So, you know, that's something to remember, I think."

She agreed and acted like we were buddies again. She said, "I think that's a very important point because I, you know, I'm working with NCIS right now. And if they didn't take their jobs seriously or, you know, they were just like, 'Oh, we're just wanting to get rid of this case,' you know, I'd be SOL. So, yeah, I understand what you're saying."

WHAT?

She continued, "And I just, I don't, I just, I didn't mean that to sound rude, like a, oh, there's no, I just, I mean, we could have like ten different performance evaluation meetings and get it to like sound a certain way, you know, and I do understand what you're saying."

NCIS helping her came out of left field and it certainly was rude. She'd not push back on lies that affected *my* job and the jobs of my colleagues. While she agreed to be the messenger of lies on behalf of one federal agency, she praised the sound ethics of another that apparently protected *her*.

I then said, "I want to be supported in my job and I'm going to look for that to happen. And I, um, yeah, this has been pretty horrible."

She said, "So, um, I'm sorry that I had to be the messenger on all of this." But where did it say she *had to* deliver lies about employees during their performance evaluations?

She said during that call that she'd talk to the union about what was happening, about employees being set up with lies. The union said she never did. And an individual in the Office of the FDA Commissioner later told me that she could change those narratives—that she *should* have changed those narratives—that she was in charge of the performance evaluations that she conducted and that she signed, and it was her responsibility to ensure they were accurate and truthful.

29 | "They're just gonna keep lying."

SHE CONFIRMED INSTITUTIONAL RETALIATION I'd first witnessed back on May 9th, 2023. Branch Chief #4 called me immediately after I'd requested a guidance document or rationale to justify his directive to use fuzzy math, something I needed to request as an FDA scientist directed to do something questionable. He unorthodoxly excluded my Supervisor #2 and said, moving forward, that I should pass things through him instead. He then ignored what I sent, effectively shutting down that work. In documents where he wasn't supposed to alter my scientific input, he newly did that. And without my knowledge, he added me to a meeting agenda to present about a topic outside my wheelhouse. If a colleague hadn't alerted me, I'd have shown up unprepared in front of my colleagues and leadership. I instead offered to query others to address that topic, and I received an award for coordinating that effort. However, he then described that effort inaccurately in a demeaning tirade he sent to my entire supervisory chain, all the way up to our Office of Science Director #2.

He'd replied to an email from me but referred to me in the third person. So, maybe I wasn't meant to see that email. I had, and it provided a glimpse behind that dingy curtain. And this was nastiness my training indicated I should immediately and directly address. So, I provided written evidence that he was lying to the same chain, but no one was brave enough to engage. My Supervisor #2, who knew he was lying, said nothing. Our Office of Science Director #2, who said we should "nip things in the bud," said nothing. They all sat by and watched a subordinate employee, while attempting to do credible work, hammered in a bout of peevish dishonesty by culpable leadership.

Our training said we should address anything like that directly, because every interaction like that could, directly or indirectly, influence the objectivity of our regulatory work. It shut down science and otherwise credible work, and that is exactly what was happening. In that petulant and dishonest email, he misrepresented work he'd agreed we should do to improve the credibility of our work, that included considerations of marketing estimates that he had a history of misrepresenting. Our CTP Director #2 wrote again and again that we needed to make those improvements. But it didn't seem they really wanted us to do that.

After that May 9th, 2023, call, I received an email from the Office of the FDA Commissioner's Anti-Harassment Program, AHP-CREW. The FDA described the policy that program was tasked with enforcing:

The FDA's Anti-Harassment and Inappropriate Conduct Policy establishes an agency-wide process under which FDA employees, contractors, fellows, trainees, and visitors shall report allegations of inappropriate conduct and harassment, sets forth manager and supervisor responsibilities to maintain a harassment-free workplace, and to take prompt and effective action when allegations of inappropriate conduct and harassment arise.

The FDA will not tolerate unlawful discrimination, retaliation, inappropriate conduct, or harassment of any kind. Through enforcement of this policy, the FDA seeks to address and correct any harassing or inappropriate conduct before it becomes severe, pervasive, or escalates to an unlawful level.

Moving forward, branch and division leadership tasked by higher-ups to address *regulatory science* issues, instead, repeatedly sent those concerns to AHP-CREW. Those issues fell outside of AHP-CREW's purview. However, I was required to cooperate each time. Over the 21 months to follow, that meant I spent a lot of time sharing information with AHP-CREW. And while those communications were generally very friendly, it was sometimes challenging to balance that repeated requirement with my workload.

In many cases, I'd quickly indicate that I didn't think what I'd experienced was harassment. Over time, however, leadership's incessant reporting to an inapplicable resource, with which I was repeatedly required to comply, created a record of something that probably was. And it was unclear why the Office of the FDA Commissioner wouldn't, instead of channeling regulatory science issues into an endless loop of ineffective reporting, ensure those issues were properly handled. It was as if they didn't really want to do that.

After that odd May 9th, 2023, call from Branch Chief #4, he reported to AHP-CREW that I'd said something. I didn't know what. I asked him and he gave me the runaround. I later learned he'd reported things to them that he'd said, but he'd apparently told them I'd said them. *He'd* talked about contentious interactions within our branch and with our division and Office of Science leadership that predated me, that he thought affected the quality and credibility of our work back then and since. As our Branch Chief #4 who took the same oath that I did, he probably should've addressed what he witnessed. He described our workplace as chronically toxic.

Regardless, I was required to cooperate. A statement and affidavit, I was asked to sign, were drafted on my behalf by AHP-CREW and an independent law firm hired by the U.S. Department of Health and Human Services. However, those documents that they prepared as part of an HHS inquiry weren't accurate. And to be fair, it might've been difficult for those uninvolved to capture events.

I was allowed to make corrections, but on a tight turnaround. And I'd have to make corrections. If not, I'd knowingly be signing something inaccurate—an inaccurate affidavit. Yikes! Imagine what it feels like to have lawyers contracted by the U.S. Department of Health and Human Services put words in your mouth in an affidavit with your name on it.

When I wrote that corrections required more time, they yielded and let me work longer. However, if I hadn't pushed back on those repeated tight turnarounds, not ample to correct or accurately represent the breadth of leadership behavior, I'd sign inaccurate or incomplete claims. And if I didn't sign, I'd be uncooperative. Their communications repeatedly indicated that I was required to cooperate, but most of those efforts weren't supported by the FDA. Our employment contract allowed only four hours of paid time for certain activities like that. Correcting just one affidavit required over 60 hours of unpaid work over just a few weeks while I also worked my normal 40 hours a week.

And they were coming on hot and heavy. My ability to summarize what they needed, while maintaining my normal 40 hours, was complicated by false claims when the threat to set up employees was increasingly realized. Those false claims also became part of that reporting. And supervisors weren't required to provide evidence with their claims. They could spout off one quick lie, and I'd be on the hook for contextualizing that lie with evidence. And that happened again and again. But that evidence didn't seem to matter.

Anyone with any other important thing going on in their life at the time all this sort of stuff happened would be overwhelmed. Staying quiet about what was happening, instead, would spare staff from what increasingly became apparent was an endless and fruitless burden. One of the last documents I provided constituted over 350 pages, organized with a table of contents and evidence that clearly indicated leadership was lying, yet CTP knowingly allowed that to continue.

In short, staff were intimidated away from reporting.

It became increasingly clear that neither CTP nor FDA wanted to really capture what was happening in CTP. For example, some of the inaccuracies written in those statements and affidavits prepared on my behalf incessantly reframed what happened. To head that off in our later communications, I'd sent clear, succinct summaries with cited documentation, direct evidence of lies and wrongdoing—not what I said happened—what our leadership did in their own words. But instead of using, for instance, those summaries or direct text or direct quotes, they'd rewrite things using their own unique twist.

For instance, they repeatedly changed a direct quote. Supervisor #6 said, "They're setting you up," but AHP-CREW responded, "So, you say they're out to get you?" Nope, no boogeyman and my hair wasn't on fire. I'd say, "No, my Supervisor #6 said, quote: 'They're setting you up.'"

Regardless, they then wrote in the document they prepared on my behalf, in my voice, "They're out to get me," and I corrected them again. Why not simply use the direct quote? For an employee to say, "They're out to get me," is very different than CTP leadership threatening an employee with, "They're setting you up," especially when AHP-CREW had first-hand evidence that leadership actually did set employees up.

Setting up employees with lies just might broach whatever FDA's threshold was—if they had one—for harassment or inappropriate behavior. It would for retaliation, considering staff were set up for asking questions they were supposed to ask that would reveal wrongdoing. So, to avoid abiding by FDA's Anti-Harassment and Inappropriate Conduct Policy, those responsible for implementing that policy could change what was said and who said it. If the FDA really wanted to function aboveboard, it had plenty of documentation, collected via their own reporting mechanisms, proving what leadership did and where corrections were needed.

It was almost as if reporting mechanisms, purportedly aimed at protecting objective science and public health, were protecting something else.

Likewise, since the fuzzy math incident, the behavior of my Supervisor #2, our Branch Chief #3 and #5, and our Division Associate Director had also changed. They delayed my work products. The timeliness of my work was a metric used to judge my performance. So, delays would affect my performance and could threaten my job. And like that new behavior by Branch Chief #4, they'd not done those sorts of things before.

It was standard practice to receive approval from my Supervisor #2, who was then also my Secondary Reviewer #1, to move work forward. For example, he'd okay work products before they were signed and finalized. Now, he sometimes wouldn't respond when asked. Some staff said leadership hoped I'd give up then be culpable for delays. So, I wouldn't drop the ball, and I didn't. And I wouldn't jump the gun and move those work products forward without approval. Doing that would affect another metric used to judge my performance.

So, I regularly reached out in writing to my Supervisor #2 to facilitate moving those products forward. After what seemed like a reasonable amount of time to expect a relevant response, I'd indicate concerns about delayed work products to my branch or division leadership, who wouldn't respond, directed me back to those who didn't, or forwarded those concerns to AHP-CREW.

And the cycle repeated.

Their behavior was documented in long email chains. So, it appeared they thought it was okay to behave like that. They delayed with games our regulatory work, a metric used by Congress to judge CTP's performance, when they could've simply cooperated. Instead, we all spent hours over months writing emails back and forth, hours that could've been used to meet our public health mission.

They, then, fumbled and bumbled their way through eye-poppingly obvious efforts to mischaracterize my health.

The position I'd applied to and accepted hadn't been remote. And I didn't want to work remotely. But COVID happened, and my Supervisor #2 said I could work from the office once COVID restrictions relaxed.

And working at home wasn't ideal. My internet connection wasn't the best and noise outside my unit was distracting. Despite efforts by employees to upgrade service on our ends, the agency struggled with coordinating a remote workforce.

We were also concerned about confidentiality. We discussed confidential business information, and I saw no evidence the agency would address, in a practical way, protecting that information in situations where employees worked remotely in shared spaces and where communications could be overheard.

So, on October 2nd, 2023, I wrote to my Supervisor #2, "My internet connection is not always reliable and activity outside my unit can be distracting. Can I occasionally work from office space outside my home?"

He and CTP's Office of Management Director #2 said the only mechanism allowing me to work from the office would be to submit a Reasonable Accommodation Request based on a medical need. But I didn't have a qualifying medical need.

The Office of Management Director #2 had been involved with discussions about the fuzzy math incident. Then she'd said, "CTP employees should never feel they are being asked to do anything unethical." So, then she'd seemed aboveboard.

I indicated that I didn't have a qualifying medical need. But in the meantime, as part of an Interim Reasonable Accommodation that my Supervisor #2 had submitted, a Wi-Fi hotspot that would allow me to work outside the home had been mailed to me. The Interim Reasonable Accommodation was meant to fill the gap until the Reasonable Accommodation determination had been made.

I asked about policies associated with use of the hotspot. My Supervisor #2 said I could work anywhere *except* the FDA office that the request, if granted, would allow.

I asked, "Anywhere?"

He said, "Yes."

"Even at a bar?" I asked. "Even at the coffee shop on the first floor of the building where the FDA office is?"

He said, "Yes, as long as you don't allow anyone to see your screen."

I then noticed on the December 4th, 2023, Interim Reasonable Accommodation form, he'd written, "A government MiFi internet device will be provided to facilitate working in locations other than your home to assist with issues of isolation that are exacerbating existing medical issues." I indicated my concern with the inaccurate statement. I retracted my application and returned the hotspot. However, the technology issues continued.

Supervisor #2 again encouraged me to submit the Reasonable Accommodation Request. When I did, he again misrepresented my health. I corrected the document and shared concerns about false representations of my health again with Supervisor #2, with the Office of Management Director #2, and with FDA's Reasonable Accommodation Office. I also indicated concerns with what seemed like an inappropriate routing of a request to work from the office through the Reasonable Accommodation process. I wrote that I was concerned a request that we knew didn't qualify would detract from requests for those with a legitimate need.

Supervisor #2 wouldn't respond about that and a lot of things. So, I asked, "If there is an explanation for your lack of responses besides retaliation for raising concerns, I need to know. The timing is consistent but I don't want to believe it is what it looks like." He didn't respond.

On December 13th, 2023, I wrote to the Reasonable Accommodation Office, "The difficulties I'm experiencing are not related to a physical or mental health disability. I'm concerned that my leadership—knowing this—would direct me here with the rationale 'to assist with issues of isolation that are exacerbating existing medical issues.'" I continued, "I have asked, but don't know what 'existing medical conditions' refers to," and Supervisor #2 or our Office of Management Director #2 never responded about that query.

At the same time, on December 14th, 2023, Supervisor #2 blamed primary reviewers for the fuzzy math incident when they'd followed leadership's directives. On December 15th, 2023, he falsely claimed "there are no definitions or duties specified for primary and secondary reviewers." And at the same time, a new laptop, that I ordered to possibly rectify connectivity issues, was waiting for me at that FDA office where I'd been told not to go. Supervisor #2 wouldn't give me permission to pick it up. He ignored that request. I alerted the Office of Management Director #2, who also ignored my request to pick it up.

I reached out to FDA's Alternative Dispute Resolution Program to request a facilitated discussion with my Supervisor #2. I hoped we could resolve these issues and move on. And our training indicated that resource could help in these sorts of situations. He did respond to that. He wrote, "I need to look into the process before I commit to participate," and he refused to participate. Instead, I completed their Conflict Coaching Training that confirmed I'd performed as expected.

In early 2024, preceding the outcome of that HHS inquiry that detailed these new questionable behaviors, Supervisor #2 was reassigned to a non-supervisory position and Branch Chief #4 left the FDA to supervise another group of Executive Branch scientists.

Our Deputy Director for Regulatory Management then took over that Reasonable Accommodation Request. He knew nothing about me, and we didn't interact during what was to be an iterative process. I don't think we've ever spoken. Regardless, he decided to put his own unique twist on my *medical issues*.

The only time I'd shared air with the man was during a February 29th, 2024, meeting at FDA's White Oak Headquarters. By then, our Office of Science Director #2 had appropriately addressed CTP's fuzzy math. Our CTP Director #2 then indicated fuzzy math was valid in his inaccurate SDR Determination. This was our only in-person Office of Science meeting, and it happened shortly after I'd provided evidence that the SDR Determination was inaccurate. I arrived for Office of Science office hours and asked our Office of Science Director #2 for his assurance we'd be doing things aboveboard. He promised we would and said, as an explanation for the shitshow we'd witnessed, that "some scientists aren't good managers."

Now it's apparent that it probably wasn't just that some scientists weren't good managers. It was probably that they weren't good scientists, and maybe not very good people.

During our discussion, our Deputy Director for Regulatory Management skulked back and forth in the background like he was trying to listen in or read our expressions. He stared at me as he walked back and forth behind our Office of Science Director #2.

Maybe he'd been "eyeballing" my health as part of another high-quality CTP assessment. Because a few days later, on March 5th, 2024, he sent his Reasonable Accommodation Determination that would be entered into the record. He wrote, "I deny the request for reasonable accommodation. You have reported that your home work environment is not conducive to working, and that you have difficulty with interpersonal interactions and feeling isolated working at home. Based on an assessment of the limited medical information that has been provided in support of your request, the Reasonable Accommodation Office, in consultation with a Federal Occupational Health Medical Consultant, has determined that you do not have a covered disability under the Rehabilitation Act of 1973/Americans with Disabilities Act Amendments Act of 2008."

I didn't say that. And how would having "difficulty with interpersonal interactions" be helped by reporting to the office?

It wasn't a surprise that the request had been denied, and that was fine. I'd said I didn't have a qualifying medical condition, so there was no medical information that would qualify. I'd *told them* there'd be no medical information that would qualify.

What wasn't fine, were these repeated mischaracterizations of my mental health. What was happening—across the board—was weird. So, I asked, "Could you please clarify what you mean by 'difficulty with interpersonal interactions and feeling isolated working at home.'"

He wrote, "I'm sorry Christy! I'm afraid I typed it too quickly. I intended to state; 'difficulty with the lack of interpersonal interactions and feeling isolated working at home.' I revised and am sending it to you now. No offense was intended in any way. Thanks!" He changed what he wrote and resent the form, but my health was again misrepresented, for the fifth time.

So, I asked him, "I'm not understanding why the reasons I provided for the Reasonable Accommodation Request have been consistently misrepresented and excluded, and I'm increasingly concerned about this. Do you know why I requested the Reasonable Accommodation?" But he didn't respond.

Typed too quickly?

Our Office of Science Director #2, who I'm pretty sure supervised him, wrote to me on April 22nd, 2024, "FWIW, I always heard and understood your request to be one about connectivity, not health issues." So, why was this allowed to continue? Why not loop him in to clear this up, to nip it in the bud? And why were so many busy high-level managers involved with not resolving my connectivity issues?

I waited a bit and hoped there'd be a correction, but our Deputy Director for Regulatory Management hadn't replied. On June 3rd, 2024, I wrote, "I'm circling back about this. The reason for the requested accommodation should be correctly documented. Will you please correct the form?"

He responded the next day, "If you feel that it misrepresents your request, please feel free to offer edits for me to consider." But when I cut and pasted verbatim text from the request, he wrote, "Thank you for your edits. Please note that I do not believe that the revisions that you provided add any greater clarity or correct a misrepresentation of your request. Therefore, I would prefer not to revise and resubmit the form."

What?

Then why ask for edits? Like when our Branch Chief #4 wrote that he *believed* it was accurate to use data combined for 32 products to review 16 individual products, now our Deputy Director for Regulatory Management *believed* his misrepresentation was more accurate than the actual text used. There was no need for his interpretation. These people seemed to think enforcement discretion was their own personal tool they could use willy nilly to construct their own narratives in every context.

So, on June 6th, 2024, I emailed our Office of Management Director #2, "I'd like to talk with you about what is happening with the documentation of this Reasonable Accommodation Request." She was our CTP Reasonable Accommodation Liaison with the FDA Reasonable Accommodation Office and had been looped in all along.

On June 10th, 2024, she wrote, "I do not recommend updating forms once the processes is completed." However, he'd already updated it once after the process was complete. He just didn't *correct* it.

This was all weird, so I gave up on correcting that form and on working from the office.

What this all indicated, again, was that leadership could misrepresent written facts, that they could blatantly lie. They'd do that in our regulatory work, and leadership would involve themselves in repeating known misrepresentations of employee performance and personal health information. CTP's Office of Science leadership could lie, and CTP's Office of Management supported that.

And that comment from our Office of Management Director #2 was interesting in a broader sense, in the context of all the regulatory documents in need of *updating*. Now coming from our Office of Management, when she said, "I do not recommend updating forms once the processes is completed," she further substantiated that CTP chose to maintain flawed records when those records created false perceptions they wanted.

Other things were revealed after those 21 months of cooperation with AHP-CREW's information gathering and engagement with other resources the FDA said protected objective science.

FDA was disingenuous about its own Anti-Harassment and Inappropriate Conduct Policy. Despite seeing evidence of retaliation and inappropriate conduct, via substantiated accounts involving numerous employees, they only ever completed that first HHS inquiry, and they saw a lot of evidence of retaliation and inappropriate conduct to follow. In that one case, they dismissed evidence of retaliation and inappropriate conduct because, they said, there was no evidence of *harassment*. Then, and moving forward, they didn't talk about other things also explicit in their policy, things like retaliation and inappropriate conduct. They only talked about considerations of harassment, and therefore predetermined the outcome of one inquiry, and precluded others. However, how could leadership indicating that leadership was setting employees up with lies while leadership set up employees with lies not be considered harassment? Apparently not according to harassment, as the FDA defined it.

When AHP-CREW saw all that evidence, so did the Office of the FDA Commissioner where that program reported. Information from the FDA Reasonable Accommodation Office was also reported to the Office of the FDA Commissioner, and they saw written evidence that CTP leadership concocted lies. FDA's Alternative Dispute Resolution Program (ADR) also reported to the Office of the FDA Commissioner. Likewise, the same evidence was shared with them.

Like the rest of us, all these people and programs were tasked with reporting concerns they learned about, within or outside their purview, explicitly to protect objective science. They, too, like the leadership that lied and retaliated, were required to report retaliation, inappropriate behavior, or any sort of wrongdoing until it had been properly addressed. However, individuals in those programs ignored what they knew or chose various tactics, like myopic considerations of harassment and parsing of incidents, to dismiss persistent patterns they saw evidence of and witnessed.

For example, ADR had been looped in for over a year. They knew staff had been threatened and that leadership was lying about employee performance. Then, in a December 19th, 2024, meeting with an ADR Supervisory Conflict Prevention and Resolution Specialist who knew all that, I said, "I have a supervisor who is making things up, in writing, counter to things that she's seen, in writing. And so, I had since provided that evidence." For example, she repeatedly lied about the Microplastics Exposures from Tobacco Products Project after seeing that 37-page report. I said, "I've provided this information. I responded to her and I provided all the documentation, and I've worked up the chain as I've been directed to do." I said, "But there is no response."

However, like others, he repeatedly reframed what the evidence indicated. He said, "It sounds like you have a difference of *perspective* on the body of work."

No, it wasn't that we disagreed with what that report *said*. She criticized me again and again for not doing the work. She was pretending that report and the mountain of evidence of that work didn't exist. And she'd responded—in writing in emails—to emails about the report, for example, asking for it as a MS Word file instead of a pdf. So, we all knew she'd seen it.

I said, "No, it is her lying. She is absolutely lying in writing about my performance, and she has linked it to my performance criteria." I said, "It is that my supervisor is lying in writing counter to evidence, in writing, that she has not only seen, but she has since been provided."

Instead of telling it like it was, he said, "So, it sounds like there's a difference, saying, I'm trying to say there's, the most *politically correct* way possible, um differences of *perspectives*."

What?

Her perspective was that she can lie. Mine, which followed our oath and training, was that she not lie. Which "perspective" did the FDA support? I said, "I've looked into this a bit, and I think what she is doing demonstrates *a lack of candor*. And I use that term carefully and explicitly because it is associated with a prohibitive personnel practice. When a supervisor is not frank, is not honest, is not, is doing what she's doing—she's concocting information."

He asked to meet again on January 14th, 2025, and said he'd met with her. He said that she was refusing to cooperate with resources meant to facilitate conflict resolution. When I asked, "What else is available?" he pointed to only the facilitated discussion that she'd refused and their Conflict Coaching Course that I'd already completed. He said, "For this situation, those are the only two things I can recommend, unfortunately."

Others had pointed to other resources and reporting requirements, such as Labor and Employee Relations, the unions, various ombudsmen, the FDA Office of Scientific Integrity, the U.S. Office of Special Counsel, and of course AHP-CREW. And ADR now knew the extent of what was happening increasingly affected our scientific integrity and probably involved prohibited personnel practices. Now there was a lot more evidence that staff were set up with lies after being told by leadership that leadership was setting staff up with lies. No doubt ADR knew, better than I, that these issues shouldn't have been talked away. Our training said we were to address issues directly with clear language. So, why did he, instead, choose to be "politically correct" to disregard a pervasive pattern of leadership lying? It was as if they wanted to allow that.

Pervasive inappropriate leadership behavior would also never be addressed when evidence was parsed. And the Office of the FDA Commissioner aided and abetted with those tactics, as well. For instance, when staff tried to discuss evidence of repeated and consistent dishonest and unethical leadership behavior, the collective of what several individuals did over time wouldn't be considered. That's often what's needed to substantiate institutional harassment or a toxic workplace. Instead, as those patterns became very obvious and probably more illegal if collectively considered, AHP-CREW said, "We are *only* talking with you *today* about *harassment* in the context of [this one interaction] with [this one person]." And I was required to cooperate with that. I couldn't force anyone to do an accurate or comprehensive or extensive or robust evaluation of anything.

And, when leadership funneled problems to AHP-CREW, they also shared their own lies with the Office of the FDA Commissioner. So, over time, AHP-CREW learned a lot about what was going on. For example, one analyst from the Office of the FDA Commissioner talked about our assessment process. She knew that every assessment in CTP's history reached findings of no significant impact and therefore facilitated marketing orders in every instance. She knew CTP scientists were increasingly pressured to create assessments that aligned with that predetermined outcome, even as that became increasingly difficult, like when information was lacking about new products sold in huge quantities. She was concerned about lead exposures that we excluded from our public disclosures.

Our Division Director lied to her. He dismissed regulatory science concerns by misrepresenting scenarios and discrediting staff. This time, he'd said our unwritten process was in writing and implied that staff requesting that written process were silly. It raised red flags when he excluded from his dishonest communications staff that could easily contradict his false claims.

The analyst knew there was no written process, and she knew that he was lying. She said, "They keep saying there's an assessment process, they keep saying there is one, I understand that, so, they're basically lying to you and they're lying to Congress." In another meeting, an analyst in the Office of the FDA Commissioner said, "He's an agent of the agency, so he's required to," then self-corrected and said, "he's *encouraged* to tell the truth."

So, there was no *requirement* of honesty from agents of the agency.

I talked with an analyst on July 3rd, 2024, just a few weeks after our CTP Director #2 had testified in front of the U.S. Senate Committee on the Judiciary. She knew that he was lying as well.

He'd said, "We have a premarket paradigm to review the scientific evidence, to understand what is in the product itself and to make a determination based on the science, whether they meet the appropriate statute by law."

He'd said, "As a scientist myself, I'm committed to following the science, and when I took this job, I vowed that and I'm continuing it now," and then confirmed, "We have an important paradigm in place."

He'd said when asked what CTP did to mitigate harms, "Well, part of it is the scientific review, to make sure that we're doing a robust review of the evidence, including the toxicologic risks and various other risks at the population as a whole."

He'd said, "It's important that we follow the science."

She said, "They're just gonna keep lying."

I said, "The problem is, for some reason, these people feel entitled to lie," and we talked about evidence that proved that CTP leadership lied.

She said about evidence I was required to provide that proved they lied, "They don't like that. They don't like that at all."

Maybe they shouldn't lie then?

I asked, "What mechanism would *leadership lying* fall under? Because it's not scientific, it's just lying. It's *behavioral*." We had the still newish Scientific Dispute Resolution (SDR) process. But our CTP Director #2 had shown he'd lie, in writing, as part of the most official parts of that process. Now he was lying to Congress, and that SDR process funneled to him. So, that process couldn't be relied on. And this wasn't a scientific dispute, it was behavioral. Of course one could argue that the fuzzy math incident, that was inappropriately handled via that SDR process, hadn't been a scientific dispute either. It was a difference of *perspectives* about if CTP *believed* math or fabricated evidence.

I said, "If [our CTP Director #2] says 32 and 16 are the same, that's not *scientific*. That's not *scientific*. That's just weird."

She was adamant, "No. That's not true. 32 is 32 and 16 is 16."

There were lots of other issues that were concerning that we all talked about, like about air emissions and babies and children exposed to things like lead. And we all often talked about the vastly inaccurate record that CTP had created by avoiding "updating" flawed documents. However, the fuzzy math incident might've been talked about the most because it was so unbelievably, incredibly stupid. It pointed to really flawed thinking and weird behavior that would be embarrassing for the agency.

And yikes! If we did that a lot, or even a little in certain contexts, that sort of dumb might really put the public health at risk.

Imagine using that math when CTP or, more broadly, FDA considered acceptable exposure levels to carcinogens or other chemicals that cause side effects. If 32 = 16 and 16 = 32, does that mean we're doubly or half as protective of the public health? Did 8 = 16 and 32 = 64, and, therefore, 8 = 64? Did 8 = 4 = 2 and 64 = 128 = 256 and therefore 2 = 256? Yikes! How far could this madness extend to FDA's regulation of not just tobacco but also to food, drugs, cosmetics, biologics, medical devices, and products that emit radiation?

FDA fuzzy math will ruin health and wellbeing and kill people. If people in power don't care about others being miserable and dying, poor health and death also affect our productivity and therefore our economy and their ability to profit from our productivity.

Anyway, with respect to where *leadership lying* might be properly handled, I asked, "Where? What FDA group would evaluate the outcomes of lying on staff in a toxic workplace? I guess that's the question. Where would that need to be heard?"

She explained, "The thing about it is, whenever there is a question of who's telling the truth and it's a supervisor and a subordinate, this is my belief, I believe that they take the supervisor's word over the employees."

So, it was coming from the Office of the FDA Commissioner, from a program tasked with investigating harassment and purportedly other important things like inappropriate behavior and retaliation, that the FDA chooses to believe supervisors over subordinate employees, even when those supervisors are lying. They'd even seen evidence that they were lying.

That means the FDA supports supervisors when they lie.

That suggests that the FDA wants supervisors to lie.

Towards what end?

And wouldn't leadership lying, again and again, qualify as harassment? What about retaliation and inappropriate conduct?

Did the FDA really want to address any problems at all?

Our still newish Office of Science Director #2 was looped in on all this and said it would change. So, I continued, "[Our Office of Science Director #2] has said there will be changes. I think it's fair for me to say, 'Okay I'm going to step back.' I'm going to say two months and just see, right? [Our Office of Science Director #2] has an open-door policy. I can go back and say, 'Hey, it's been two months. Do you think that there still will be positive things moving forward?' That's the approach that I would like to take."

I'd just been told by the Office of the FDA Commissioner that leadership "didn't like that at all" when I'd cooperated with required reporting that proved they lied. So, in hindsight, I think I was feeling a little intimidated. And in hindsight, and knowing what she knew, I don't know if someone in her position should've said, "They don't like that. They don't like that at all," not if the FDA really wanted employees to report.

I said, "I don't want to cover anything up. I want FDA to be able to do what it's mandated to do. Despite what [our CTP Director #2] said to Congress, [he] knows there are problems and I want them to have the opportunity to fix things in a way that doesn't completely destroy our mandate. And it's also a huge burden on me, quite honestly, you know, I don't get paid what they get paid. I want them to do it. I want to give them the opportunity to do their jobs."

In the meantime, Supervisor #6 had said, on July 30th, 2024, that complicit leadership was setting up employees by lying about their performance. So, it did seem "they didn't like that" and that "they take the supervisor's word." When I talked again with that analyst, on August 2nd, 2024, about that and other problems in CTP, I said, "[Our Office of Science Director #2] knows about it. He said he's trying to change the culture. And so I do think it's fair, like we talked [about] before, to provide him that time. But at a certain time, the clock's going to run out. And I need to decide for me when that happens and when I do take it further. I feel that I have communicated enough to the right places where I am not not following my obligations, where I'm not helping bury a problem."

She agreed, "Right, you have exposed it, for lack of a better word, you have made everyone aware."

Leadership then carried through on their threats with even more blatant false claims that escalated in severity. When we talked about that on August 29th, 2024,

I said, "This sort of thing is going to prevent us from speaking up in our jobs." I said, "This is and it will affect the science that we do."

And she agreed, "Yeah."

So, the Office of the FDA Commissioner saw repeated evidence that supervisors promoted lies in our regulatory science work. They saw evidence that supervisors lied in our work, lied about our work, lied about their behavior, and lied about employees. They and the U.S. Department of Health and Human Services knew it was happening in the CTP Office of Science and in our division, since at least 2019, and chose not to correct it. The public health concerns associated with all the very serious cascading outcomes of those behaviors were ignored. And moving forward, those behaviors weren't only ignored, they were supported.

30 | More Hindsight

ON JULY 31ST, 2024, the day after my Supervisor #6 said, "They're setting you up," she started our next meeting saying, "It's my understanding that, I think there's only been like one or two cases, with NEPA, in general, where it hasn't been a FONSI. Like, it's basically, I've heard it described as something *without teeth*, like it's basically *administrative*."

What? Heard? Where?

I checked my watch to see if I'd mistakenly woken up on Mars.

I looked outside and saw no pigs flying.

She was saying that since NEPA was signed into law in 1970, its implementation reached a finding of no significant impact in every case, but one or two. That wasn't true, and it was unclear why she started our meeting saying that. She had the internet too and could look that up. Was she trying to convince me of that? Did she think she could?

And when I asked, "You mean within CTP?" she confirmed, "Uh, broadly, broadly." She meant "broadly."

This was coming from a supervisor in a branch of over two dozen scientists dedicated to implementing NEPA, and it was an astonishingly misinformed thing to say. For example, instead of FONSIs, 221 environmental impact statements were published in 2024, 162 in 2023, 185 in 2022, 186 in 2021, 254 in 2020, 219 in 2019, 323 in 2018, 257 in 2017, 312 in 2016, and 381 in 2015.

No, Christy, just "one or two cases" in the history of the world.

What was she getting at? If that was true, did that mean the precedent was a FONSI and therefore that was what we were expected to support? Why not just put that and our policies in writing to prevent all this weird chatter?

By indicating it was "basically administrative," she seemed to indicate our implementation of that law was performative and amounted to checking boxes, donkey work, sausage-making, or rubber-stamping, maybe? This sounded like the "Why bother?" argument. And I'd heard some CTP leadership and staff say almost the same thing, word for word, to disregard concerns about previous approaches that weren't credible and to avoid efforts, going forward, that were.

CTP hadn't had two cases, or one case, or any cases where there hadn't been a FONSI, so she really believed or wanted to tell me that since NEPA was signed into law, in 1970, its implementation reached a finding of no significant impact in every case, but one or two.

FDA spent thousands for training—for each of us—that meant we all knew all these sorts of claims were ludicrous. And regardless how any individual described or felt about what we did, we needed to follow the law. I'd not read that because some people thought NEPA was "basically administrative" or "without teeth" that we at CTP could waste millions of dollars annually—in our branch alone—to lie and rubber-stamp in lieu of doing credible work.

Our training and ubiquitous external messaging said our process was "to ascertain the environmental consequences of [our agency's actions] on the quality of the human environment and to ensure that the interested and affected public is appropriately informed" as "an integral part of its decision-making process." Our Division Director had said, "An assessment is a stand-alone document for the public to understand the government's environmental considerations." He said, "It's recommended that an assessment include discussion on the following topics," with "air quality" first on his list. He said, "Potential impacts of tobacco products may include impacts resulting from secondhand and thirdhand exposure from use." He said, "It's very important to compare the health risk of a new product to comparative products that are likely to be consumed by users of the new product." He said, "FDA's goal in product regulation is to reduce the public health risk and individual health risk to the user posed by tobacco products available on the U.S. market."

So, what was *she* saying? She was trying to handle things using that *esoteric* unwritten word of mouth that she'd complained about, and she knew better.

I said, "Regardless, there should be procedural documents that advise us how to reach those findings. And as a NEPA practitioner, it's my responsibility, because I sign my name to it, to understand our process, and that is what I've asked for." I said, "I know that NEPA is a real law."

She then shapeshifted and agreed, "Yeah."

I continued, "Even if there is a finding of no significant impact, the point is to inform the public about the decision-making process. So, it's not about the outcome [referring to the federal action], it's about the information. It's saying, 'Okay, this is the information that the agency used to come to that determination.'

And the public, this is for the *public* to understand. So, [the implementation of] any law is procedural. I don't really—"

She agreed, "Right."

I continued, "I mean, so, yeah, um yeah, it's about doing that accurately."

"I know in Toxicology [Branches in our division]," she said, "we generally have to go by precedent, like, what's come before." So there, as well, we couldn't use the best available, most up-to-date science.

I said, "If we learn about something new, we do not have—unless there is some justification that needs to be included in our procedural documents—we do not have grounds to exclude things that we know about." I continued, "One thing that is happening, is that a previous reviewer doesn't do something, that means it doesn't happen in a subsequent review. But sometimes it's just simply missed. And there hasn't been, and I'm, you know, I'm not going to say any of this without being solid and not without understanding the importance of what I'm saying, some of our reviewers are sloppy. They just don't care, and this is in Environmental—"

She interrupted, "Yeah, I've seen that, [it's] not just Environmental [Science]." And she said the lacking scientific basis for our regulatory science work explained disturbing leadership behavior. She said, "I think having like a *scientific basis* for things might be what's underlying a lot of this, from what I'm seeing."

She also said, "My feeling on it is nobody else knows. I feel the keeper of this knowledge [about how we do things] is basically [our Branch Chief #3 and #5], in terms of who's been here and has made decisions on these types of things. And I feel like probably the rest of our division, the rest of CTP, like, it's just like, 'I don't know,' like, 'What is this?'"

In hindsight, I see how regularly switching supervisors prevented those moving in and out of these positions from recognizing patterns. It had taken me some time, as well. And keeping us siloed also prevented that. However, Supervisor #6 came from another branch in my same division where the same behaviors were applied by some of the same people. So, she was picking up on things quickly. I'd previously interacted with her a bit on division-wide work, and she came across as smart, dedicated, and conscientious. Now, she was swinging back and forth between credible considerations and sausage-making.

I'd, too, initially thought mid-level management might not have understood. So, I said, "I think in the beginning that was the case, but there have been enough conversations and there's been enough explanation provided." I said, "If you look at my correspondence, I say, 'Please reach out with any questions. I'm happy to provide context. We can go over things together.' They just won't initiate. So, at a certain point, I think, you know, I think a lot of us said, 'They don't understand.' I think [our Division Associate Director] understands well enough to avoid things. He, if he truly didn't understand—"

"He's writing a book!" she exclaimed.

We'd previously taken the bait, writing off their frazzled bumbling as unintentional incompetence that is still not okay. However, their performances elsewhere were far less bumbling. For example, our Division Associate Director acted like he didn't understand things he'd previously described and directed, in detail. He complained that his workload didn't allow him time to understand things he'd already indicated he understood. And at the same time, he'd been able to write a 238-page, highly technical toxicology book that proved he understood the inadequacy of the work we did. We'd just learned this and the shock reverberated through our division.

He'd duped us. He wasn't an idiot. He played one when it was convenient.

Our Deputy Director for Regulatory Management attributed his own incompetence to typing too fast. He said small text precluded any public health consideration whatsoever of a tobacco product by the public agency tasked with doing that. And he wrote as if he hadn't known about the 117-page white paper that was developed and communicated where he worked when it was produced and shared. He feigned not knowing about concerns about lead after spending decades working with metal contaminants in products regulated by the FDA.

Our Branch Chief #3 and #5, on the other hand, feigned knowledge. She was a gaslighter. She chronically misrepresented the science and manipulated arguments with weasel wording, like framing a data quality issue as an editorial comment, then administrative issue, then a legal argument. She said we didn't consider product use and human health when our training and publicly posted documents and communications showed we did. And she brushed off concerns with microplastic pollution because, she said, "Everything in our life has microplastics, or produces microplastics in general."

These were the people that designed and enforced our approaches. Our new CTP Director #2 and our new Office of Science Director #2 certainty had plausible deniability when they were new. But by this time, they knew what was going on. But what had mid-level management told them?

Anyway, during that July 31st, 2024, meeting with Supervisor #6, I said, "It's really interesting to see, [our Office of Science Director #2]'s been receptive in my sense, and I don't know this for sure, but my sense is he's learning. He's seeing what's happening. He still has the open-door policy, and he must get all sorts of stuff that filters up to his level. And he asks questions. The second, you know, a problem came about, he sat down. I put together a PowerPoint. He went through it. He asked questions. He asked the right questions, which showed me he was interested in engaging. [Our Division Associate Director] and [Division Director] do not do that."

She responded, "Yeah."

As leadership became less responsive and increasingly petulant, our Office of Science Director #2 repeatedly encouraged engaging, for instance, writing directly to me, even after I'd been told I was being set up, "All I can suggest is persistence." Why would leadership below him punish staff for raising concerns when he continually asked us to do that? If my Office of Science Director #2 directly wrote to me that I should be persistent, I needed to do that.

She also said, "My takeaway from the mid-cycle [performance evaluation] was that it was like a warning sort of thing, and that's kind of what I was trying to, I'm trying to think about ways to protect yourself from this type of, you know, retaliatory, or misunderstanding, or whatever, whatever this is, like, I'm trying to understand it better." And she provided advice about how to document things.

She was upset about the expectation that she must deliver lies about staff performance in lieu of our Division Associate Director, who sent her those lies to deliver. She said, "I think, it's kind of like, why, why me not [our Division Associate Director]?" She said, "It's basically, 'Okay, here's, here's what you say,' and then like, 'Okay.'"

I responded, "So, then, that's an interesting part, too, is that you've been my supervisor for a while, and some of the things he wrote, it's not true based on what you experienced. So, that's a little strange to me as well." For example, he'd falsely represented my performance on review-related work that happened when she was my supervisor. I said, "The *only* review I had was when *you* were my supervisor and that was not delayed. And it was done, again, it was signed early."

She agreed, "Yeah."

I said, "How is it that he describes events when he wasn't [my supervisor]?"

She said, "Right, and I will say, I talked to some other people on the team, and they just made observations that some of this sounds like it's coming from [our Branch Chief #3 and #5]." She was referring to the fact that other scientists on Team 1 were being lied about. They, too, were warned that management would lie about them during their mid-year performance evaluations. We were being warned that management was empowered to lie as part of a process that affected our jobs, our means to pay our bills. So, this implied we'd be left alone and might keep our jobs if we give in. She then referred to our Branch Chief #3 and #5 as "the keeper of our branch knowledge." She certainly did know where the bodies were buried, but that didn't mean her supervisors were off the hook.

Strangely, and maybe because we both believed Executive Branch employees really couldn't act this way, we were still hopeful. She said, "What I'm hoping to do, is eventually we can do these Team 1 meetings and then maybe bring it up during a branch meeting. Because I do think that there might be an opportunity for change right now." She asked, "Can you do me a favor? Can you do me a favor though before our next team meeting?"

I said, "Yeah."

She asked, "Can you put together your list of questions from the emails you have?" She'd seen the history of issues and how they'd been obfuscated. Yet, she wanted me to develop that list of questions right after she'd told me I was being set up for asking too many questions.

She was my supervisor, so I said, "Okay."

She continued, "Because I think it might be, I just want to have, like have them, if you don't mind, like have them all together so that we can, like, we can like brainstorm as a team and think about approaches and stuff."

"Okay. I will do that. All right," I replied.

She continued, "Yeah, because I don't know. I feel like you shouldn't be the only one like sticking your neck out here. So, if we can get like a branch-wide thing that would be really nice."

"Yeah. That sounds good. We can try." I said, "Yeah."

She continued, "Because I've also heard other people on the team say the same things. So, I know it's not, you know it's not just you." She referred to Team 1. But some on Team 2 and Team 3 also raised the same, similar, and additional concerns. A Team 3 colleague frequently repeated, "We straddle a line between what is legal and what is illegal." And it was supervisors as well, even Supervisor #2 and Branch Chief #4 lamented the unethical machine we were part of.

Anyway, I replied, "I think it makes sense to start at the team. That's a good idea. I like that."

She said, "Okay, we'll try to figure something out. And let me know if you want to change your mind and stuff because I want to be able to, I just want to make sure there's not a retaliatory thing and I want to help as best I can. And I think the way to do that would be to get like a consensus starting with our team, um, stuff like that. But, you know, I'm open to ideas."

I said, "What was written," by our Division Associate Director, "that is retaliatory." But I said, "I think we can move forward with what you're talking about, because the idea is to make improvements. And if we can move forward and not have it happen again, that would be awesome. So, I like your approach. That works for me too."

She replied, "All right. Thank you so much."

August 1st, 2024

Supervisor #6 had been consistently pretty professional and generally quite pleasant during that August 1st, 2024, meeting where we talked about my new assignment with those 14 oral pouch products. We'd talked for over 30 minutes with my Secondary Reviewer #3. We talked about the great microplastics work they said I did and how that would come in handy to ensure we used the best, most up-to-date information in the context of that assignment.

There'd previously been no problem communicating with colleagues doing parallel work, or with our branch chiefs. We did that to try to stay informed and be consistent. I'd even confirmed with our Division Associate Director during that April 24th, 2024, meeting that any sort of communication like that was encouraged. We'd subsequently been told we needed to work independently and take the initiative to move things forward by the Director of the Division of Regulatory Project Management.

We weren't expected to ask permission to communicate.

However, Supervisor #6 instead asked us to let her ask if we could communicate with colleagues doing parallel work and if secondary reviewers could communicate with our Branch Chief #7. She said, "I don't want you guys to stick your neck out for stuff. I'll run it by and ask if secondaries can communicate directly with her, or, basically how secondaries are going to communicate." She said, "I think that would also really help with the feeling of silos as well, because I know everyone's feeling like pretty isolated."

August 2nd, 2024

The next day, I sent her the list of questions she'd requested on July 31st, 2024:

Hi [Supervisor #6],

Thanks for the chat about this earlier this week. I need to focus on my review and wanted to get this to you asap.

There's a long, documented history of staff attempts to obtain critical information. That history can be shared, and as has been always offered, clarifications will happily be provided.

As requested for our August 8th meeting, below are several of those critical, outstanding requests. These questions remain unanswered, and staff have not been told that this information is not needed to do our work:

1. Are assessment preparers and reviewers to agree with the argument presented in the assessment? The argument in every case is that there are no environmental impacts based on the information included in the assessment. Many reviewers have indicated concerns with that information and with information that is excluded. Requirements and messaging increasingly indicate that we cannot rely on the information used to support subsequent findings, e.g., data provided by an industry repeatedly proven to mislead governments are not validated, information is not gathered and considered systematically, and the robust, comprehensive scientific approaches that [our CTP Director #2] *told the Senate Judiciary Committee in June that we use are not used.*

2. How have FONSI determinations been supported by the information included in assessments? Information included in assessments varies and it is unclear if/how the FONSI determination is made.

3. As per FDA and CEQ requirements that we use valid approaches, when and how will data that we use in our assessments that is provided by applicants be validated? Until then, such information, and other information that does not align with FDA and CTP policies, should be excluded. When we exclude that information, how will FONSI determinations be supported? If that information has historically informed the FONSI decision, and CEQ guidance indicates that it has, how should reviewers compensate for those changes? (This need is part of the reason we also need clarify about #2.)

4. Moving forward, when we remove information from our assessments that is not based on a comprehensive consideration of available information and that has not been substantiated, what priority information will our branch focus on substantiating that can be included in our assessments? The effort needed to prioritize and address such topics is beyond the abilities of the individual primary reviewer, considering expertise, workload, and our need to be consistent. My understanding is that many staff would be excited to participate in any such efforts. The scoping effort restarted a few times seems like it might get at this need.

5. How are reviewers to perform scientific reviews and assessments without using or considering thresholds? We have received this directive in writing from [Branch Chief #3 and #5], [Branch Chief #4], *and from* [our Division Associate Director]. *Additionally,* [the Director of the Division of Regulatory Project Management] *recently wrote that we do not use thresholds in CTP. However, we do* [some did], *and the scientific nature of our work requires that we use or consider qualitative and quantitative thresholds, especially as new information is learned. This messaging is disparate with our training, messaging, and practices. Is it an FDA or CTP policy not to use or consider thresholds, and how is any such policy justified?*

6. Reviewers are increasingly instructed, in the absence of a consistent approach, to use independent judgement when we come from very different backgrounds. Yet, other messaging indicates we are to be consistent. Additionally, scientific decisions in general require a comprehensive consideration of available information, a task relative to each topic that can require teams of scientists many years to develop. FDA-wide, we are to base our regulatory decisions on the comprehensive body of information, but in CTP individual reviewers are being asked to independently develop assessments of a broad set of environmental parameters about which there is substantially less available information and considerable uncertainty. How will CTP reconcile such disparate assessments?

7. What is our one, current and consistent set of processes? Links and general mention of such information, e.g., in emails from leadership, do not include information about a consistent process and do not address important questions that we are encouraged to ask. In the future, could leadership indicate specifically where in the referenced documents answers to important questions are available or answer the questions directly? Some are yes-or-no questions.

8. Provide a contact to facilitate access to CTP's National Environmental Policy Act (NEPA) Procedural Documents and any other administrative record or file associated with our NEPA compliance. This need is additionally critical because [the Director of the Division of Regulatory Project Management] *recently indicated (below) that we are not to follow NEPA and made other assertions that have confused staff. It is critical that Regulatory Health Project Managers understand our branch's purview and that our branch staff know if we are not following NEPA.* [The Director of the Division of Regulatory Project Management] *also directed me to review over 26 million regulatory documents, when the foundational documents that I'm requesting would not require such an effort. The agency has supported recent NEPA training for some of our branch staff and those staff in particular (but all staff) should be looped in about compliance—we are told that we are to understand—that is part of our responsibility, but we aren't provided access to our procedural documents. We understand that the agency has historically decided how to implement NEPA, and policy entitles staff to see such documents. 21 CFR 10.70 provides that access.*

9. Is the solution developed by [our Office of Science Director #2] *that precludes previous misrepresentations of marketing estimates our approach moving forward?* [Our Division Associate Director] *has indicated that he supports the previous practice and has refused to respond in follow-up communications suggested by* [our Office of Science Director #2] *to clarify if we are to revert to that practice moving forward. Our branch needs to understand which approach is to be used and needs clarity about any requirement that reviewers misrepresent data moving forward.*

There are certainly other concerns, but I've tried to distill the most pressing from my perspective based on our regulatory purview. I believe others will have different input from their perspectives. [Supervisor #4] *was tasked while our supervisor with providing some of these answers but did not circle back, and I can forward those communications if needed. I'm happy with you forwarding this to* [Supervisor #4], [our Branch Chief #7], *or any others if that helps.*

Again, I'm happy to clear up any apparent ambiguity in what I've communicated here. These issues can be complex, and I do my best to provide the crux of the issue, but I realize despite the extra time needed, that reiterating explanations is sometimes needed.

Have a great weekend and see you next week!

Christy

August 6th, 2024

She was again consistently pretty professional and generally quite pleasant during the next meeting I had with her that also lasted a bit over 30 minutes. She'd invited two other colleagues and talked about the Issue Tracker. One of those colleagues worked on the Issue Tracker and on the other effort our Branch Chief #7 falsely claimed was the same, and that therefore meant the Issue Tracker that would truly ensure consistency, and therefore accountability, was unnecessary. That colleague confirmed that the other effort was not the same.

Supervisor #6 said about our Branch Chief #7, "I guess she's starting up the [other] Issue Tracker group again, and that's why I wanted to hold this meeting, because I'm trying to figure out, I don't know if [the other colleague] has information on that or not. I'm not sure what's going on."

That colleague said that the other project "was started last year and there's been no updates until now," while there'd been consistent progress on the Issue Tracker we were working on.

Supervisor #6 asked, "Is that the same thing as what you guys are working on [on the Issue Tracker]? Because I put on the [branch] meeting [agenda] that we're doing next week, the 'Issue Tracker Discussion,' and then [our Branch Chief #7] contacted me, 'Oh, we're already doing the Issue Tracker.' So, then I was like, 'Well, what's?'"

The colleague confirmed, "They're different. They're totally different," and continued, "The issue *summary* document that we're working on is basically just request language. That's all it is."

Supervisor #6 said, "This could be worth bringing up in the branch meeting if it's different. I just wanted to make sure you guys weren't working on something that was already being worked on."

It was great to hear we'd talk about the Issue Tracker at an upcoming branch meeting. So, I said, "Yeah, thanks, that's good!" And I further clarified, "It actually incorporates the other project."

Efforts like this were usually presented to the branch. So, I'd already prepared that presentation. Before that, I'd also already shared a summary document with my branch chiefs and supervisors that described the Issue Tracker. During that August 6th, 2024, meeting, I'd shared my screen and ran through those details. That summary document clearly showed how that other effort had been incorporated into the Issue Tracker. They'd sent me their draft request language, and I'd added it alongside a note that any updates they made would be reflected in the Issue Tracker. That was explicit, and there was no knowledge of similar summary or presentation documents associated with that other effort, just that request language they'd sent me.

I shared my screen and showed where, "In here, it includes the background, the need, the solution, and then as part of that, it includes in Section 6.2, the draft request language. And so that's the project that [the colleague] is talking about. This [Section 6.2] would be replaced with whatever updated product comes out of the other group."

Supervisor #6 responded, "Perfect."

I was glad the other group would soon provide their final request language so we could update Section 6.2.

So, I said, "The other effort that you're referring to is incorporated in this. So, it's cool actually to have both of them going on at the same time."

She said, "Hopefully you can do a presentation on it, if you feel like you're ready, at a branch meeting."

I replied, "I think we will be. Thank you so much!"

The other colleague she'd invited previously worked for a different FDA center and they talked about how what we did in CTP differed from other FDA centers. He talked about the importance of what something like the Issue Tracker would provide, to ensure consistent work and improve efficiencies, like memos used in the other FDA center where he worked but that we didn't have. He also talked about other established means primary reviewers there had to raise concerns and document considerations of those concerns that we didn't use.

CTP was purportedly held to the same ethical and scientific standards applicable to all of FDA, and all of science. But we didn't implement mechanisms to ensure or improve consistencies, efficiencies, or accountability. However, because some scientists often inherently look to evidentiary standards, many of us looked for means to legitimize our work, to bring it aboveboard. The Issue Tracker was one of those efforts I'd been encouraged by Supervisor #2 and our Division Associate Director to develop further, after our CTP Director #2 and

Office of Science Director #2 said we'd see improvements following events associated with the fuzzy math incident. However, it was unclear why we didn't just use mechanisms already established in other FDA centers where many CTP leadership previously worked.

Discussion of the Issue Tracker was subsequently removed from our meeting agenda, and, on October 23rd, 2024, Supervisor #7 oddly said about the Issue Tracker, "The branch will not bless this project." I pictured the pope standing over a dossier of projects, blessing those incapable of creating accountability and excommunicating the rest to a dumpster behind the altar.

August 8th, 2024

On August 8th, 2024, Supervisor #6 was consistently pretty unprofessional and generally rather unpleasant during a Team 1 meeting. She started that call discernibly discombobulated. She asked that I resend the list of questions she'd requested, despite having said, "They're setting you up," because "you're asking too many questions." She mentioned receiving it but couldn't find it. I sent it again, but she never talked about it. And she never talked about any of the input she'd repeatedly requested from our team of scientists that she supervised. That had been the plan for that meeting she'd scheduled for that day.

Instead, she spent the entirety of that meeting talking about inappropriate conduct in our branch. She talked about her interactions with and observations of branch leadership. She said Team 1 scientists were targeted by branch leadership and colleagues in our branch. She said Team 1 scientists were being "left out" of work opportunities. And she indicated that review teams were gerrymandered.

She whipped back and forth between purportedly doing us a favor by telling us that and rudely delivered warnings to change our behavior. She wouldn't say *what behavior.* I knew of no behavior by Team 1 scientists that violated policies or contradicted our training.

This was like she'd done before, when she'd agreed to be what she characterized as "the messenger" of division leadership's lies, when she said they were setting us up with lies. Then and now, she was distraught. Then, and again during that August 8th, 2024, meeting, she seemed more upset about what she'd agreed to do—and the hardship that *she* was experiencing—than what was being done. She claimed martyr status, as if her decision to inform us was a burden to her. But *she chose* to be complicit.

Watching one after another supervisor flop back and forth via their own existential crises while they intermittently freebased the cool aid got old. Over time, however, it demonstrated a standard set of inappropriate behaviors across leadership and management. And we purportedly still had policies that our training referenced. So, we were patient and held to our convictions while she flipped and flopped. Each time she said something weird, we calmly asked her to provide clarification and report the inappropriate conduct she said she witnessed.

Instead, she repeatedly indicated that if we changed our behavior, in some undescribed way, that we might not be "left out." She said, even when those unarticulated criticisms of our undescribed behaviors were "unwarranted," that she wanted us to know that we were being criticized, so we should change our behavior. She simply wouldn't say what behavior, and delivered again and again, while she shifted from friendly to dismissive to peevish, that same sentiment—conform to unwritten and undescribed norms or be "left out."

We'd ask, "Which behaviors?" and she'd thrash around a bit, like a possessed cat spritzed with holy water. Instead, she'd grumble, "Maybe I shouldn't have said anything," and we'd patiently ask again.

We'd need to know what behaviors needed to stop to prevent being targeted, especially those behaviors where she said criticisms were "unwarranted." We didn't know which of those behaviors, that followed agency policy, needed to cease so we'd be welcomed into the fray. And if we changed those behaviors to behaviors that warranted criticisms, we'd be complicit.

But she'd never discuss those behaviors.

What she eventually described indicated that review teams for some tobacco products were being loaded by primary reviewers and secondary reviewers from Team 2. Assignments were supposed to rotate among teams and among our swath of scientists to ensure equal work opportunities, and just as importantly, to prevent the same sort of silos where wrongdoing happened previously. Ideally, mixing up teams of scientists created checks and balances to help ensure we did credible work. What she told us indicated that wasn't happening.

We asked if she'd reported that and she said, "No," because she said she was "respecting their decision" to do that.

What? That was her explanation, that she was "respecting their decision."

Who? She indicated it was branch leadership and Team 2. Branch leadership at that time was Branch Chief #7, the Team 3 Supervisor #4, and Branch Chief #3 and #5 who'd returned to her role as the Team 2 Supervisor #3. Our Branch Chief #7 had been promoted from Team 2 and reported to our Division Associate Director.

She said she was helping accommodate those decisions by agreeing not to assign Team 1 scientists to review some products. So, certain applications were packed with primary reviewers and secondary reviewers complicit in questionable practices and she said, "I'm respecting their decision."

A few months earlier, she was concerned about wrongdoing and retaliation. Since, she'd provided additional accounts of the same behaviors. However, by August 8th, 2024, her radicalization had taken hold and she'd decided she'd play along. To stave off being targeted and "left out" of work opportunities, she said we'd need to change undescribed behaviors in undescribed ways.

What message does it send when a supervisor tells employees they're being set up, lied about, targeted, left out of work opportunities—that she won't report it—and then she decides to play along? What knock on effects does that sort of behavior have? We asked what we could do, and we asked who it was coming from, from all of our colleagues?

She said, "I don't have an answer for that, and I'm sorry that I don't. I wanted you to have this information because I want you guys to be aware of what's going on, and I'm, I don't know what to say. I don't want to share this information with you. It would be a lot easier for me to simply be like, 'Oh, you guys are doing great. Nobody has any problems. Everything's fine.' Right? Then I've done my job. I'm done, right? I'm trying to tell you that there are [problems], and maybe it is just Team 2."

Several of us asked, "Why?" this was happening.

She said, "I think that it is important, I mean, you guys have raised important issues, and either it's not being heard by the leadership or there are certain people in leadership that have an attitude that's being handed down to the rest of their team, or something like that."

During that call, she did eventually make one specific criticism, not of any behavior by any of us, but of one specific effort. The only concrete criticism she shared during that call was about that Microplastics Exposures from Tobacco Products Project I'd led since 2021. She said that Branch Chief #7 said there'd been no activity on that project—since 2022. And now *she* seemed to agree, "There hasn't been anything, has there?" She asked, "How is that project going?"

In my head, I reflected on evidence proving that everyone knew.

Supervisor #6 would've certainly immediately dispelled that apparent misconception by our Branch Chief #7. We'd had detailed conversations about that project very recently and over the preceding months. I'd given her updates. I'd looped her in on emails. I'd shared evidence of progress with her. She'd even said the Office of Science had recently prioritized that work. The process that got that project listed there involved summaries and updates that I provided to branch leadership, and that were subsequently forwarded and reviewed by division leadership before being forwarded and reviewed by Office of Science leadership. As part of that process, I'd answered follow-up questions and provided clarifications. I'd even recently received an award for that work from the Office of the FDA Commissioner that had been included in our division newsletter. And I was helping coordinate a meeting across all FDA centers that our Commissioner would speak at. That meeting at FDA's White Oak Headquarters was a great success!

And, on August 1st, 2024, only seven days earlier, she'd pinpointed that project when she referred to the value that my microplastics work and expertise would provide to CTP's first consideration of oral pouch products containing nicotine extracted from tobacco.

During that August 8th, 2024, meeting, I indicated work on that project was going well, but she knew that. I said, "It's fine," and offered additional information to our Branch Chief #7. I said, "I'm absolutely happy to do that. I have reported my progress and activities with microplastics and kept the group in touch." I also told her that the division had received updates as well.

She said, "Okay."

This had to have been a misunderstanding. There was a tremendous amount of written evidence known to the entire branch and beyond. Supervisor #2 had been a participant on that project for two years, and I'd provided activity updates to all my subsequent supervisors. I'd written and uploaded excruciatingly detailed meeting minutes to our branch's shared folders. I pointed to those documents and other relevant resources in emails and presentations when I provided updates about plastic polymers in tobacco products, related agency and government-wide efforts, and associated legislative updates to the branch.

As that project progressed, I had calls about various aspects of that work with our Division Associate Director and Branch Chief #3 and #5. I regularly alerted the division about issues that might fall under our regulatory purview, for instance, learned via three different meetings I attended monthly for two years. In fact, I'd *just* emailed branch, division, and Office of Science leadership about several issues. Those regular emails went to our branch leadership and to our Division Associate Director and Division Director. And for several years, I'd worked with our Branch Chief #7 on a life-cycle assessment project that she led, and we frequently talked about sourcing information about tobacco product materials, to serve both projects. Most recently, I'd been working with the CTP Office of the Center Director, for the FDA on behalf of CTP, to review pending treaties and legislation with respect to plastic polymers and microplastics.

And there was more! That project was awesome!

So, this had to have been a misunderstanding.

August 13th, 2024, until March 7th, 2025

A week later, on August 13th, 2024, Supervisor #6 made dishonest claims about Team 1 scientists. She claimed we'd violated the U.S. Department of Health and Human Services' Code of Conduct. She wrote, "I want to address your behavior from last week's meeting. Specifically, during Thursday's team meeting you and two other members of the team raised your voice to me in an inappropriate manner and stated, to the effect, that the reasons for the team's issues were because of my poor supervisory skills. Additionally, you made negative comments about the Branch Chief's leadership and made rude remarks about other members of the branch, such as stating that they were not doing their jobs."

She continued, "I consider this unprofessional and inappropriate, which led me to end the meeting early as I felt uncomfortable. Your conduct violated the HHS Code of Conduct, specifically, "actions which deny the dignity of individuals or conduct which is disrespectful to others must be avoided."

Then she stressed, "I want to stress that this type of behavior is unacceptable and will not be tolerated. Should this conduct persist, it may result in disciplinary action in the future."

And she signed her email, "Kind regards," but she was lying.

Now we knew she'd accepted the assignment to set up scientists with lies. Like the lies she said she'd received from our Division Associate Director, she now posed dishonest arguments lacking details to substantiate her claims. Despite her tantrums that she was "being thrown under the bus," she was now behind the wheel blowing through stop signs.

Advice she'd just given to protect ourselves against the very same behavior—to document everything—documented that she was lying. The recording of that August 8th, 2024, meeting proved she was lying, and she'd need to explain why she made false accusations about CTP scientists. It also again substantiated that employees had again been threatened by leadership, now by Supervisor #6 who indicated wrongdoing in our division hadn't been corrected.

When I received her email that day, I replied, "Good morning [Supervisor #6], Your accusations are false. Sincerely, Christy," but she didn't respond.

On August 15th, 2024, I wrote, "Good morning [Supervisor #6], would you be willing to share your notes from our August 8th, 2024, Team 1 meeting? It is unclear how you reached the conclusion that you did and your accusations are serious. Thank you, Christy."

She didn't respond.

On August 30th, 2024, I wrote, "Hi [Supervisor #6], please let me know if you are willing to share your notes from our August 8th, 2024, Team 1 meeting. It is unclear how you reached the conclusion that you did and your accusations are serious. I am certain that I am not guilty of the claims that you've used to characterize my behavior as violating the HHS Code of Conduct. Making such false accusations can impact my career and my livelihood, and it creates a culture where employees feel threatened and unable to speak, which will effectively lead to censorship. Thank you, Christy."

She didn't respond.

On September 11th, I wrote, "Good morning [Supervisor #6], please let me know if you are willing to share your notes from our August 8th, 2024, Team 1 meeting. Thank you, Christy."

She didn't respond.

Others reported what she said and did, but no one would engage. For instance, colleagues said she cancelled meetings and wouldn't engage again. Our Branch Chief #7 was unresponsive, so communications moved to our Division Director, who wrote, "With regard to [Supervisor #6]'s notes, these are hers alone and will not be provided to you," and he wouldn't engage when provided evidence and detailed accounts that proved she lied. Likewise, Office of Science Director #2, CTP Director #2, AHP-CREW, and other contacts and resources in place to ensure FDA scientists could do objective work ignored and allowed these escalating behaviors into 2025.

She was purportedly obliged to provide details and documentation when she claimed employees violated policies. Instead, they protected a supervisor they knew was lying. They refused to engage when scientists were set up after they were told they were being set up and targeted by branch and division leadership. *Their* conduct violated the HHS Code of Conduct, specifically, "actions which deny the dignity of individuals or conduct which is disrespectful to others must be avoided."

Indiscretions by and described by Supervisor #6 during that August 8th, 2024, meeting, and previously, were ignored, then continued. The analyst from the Office of the FDA Commissioner had said that "whenever there is a question of who's telling the truth, and it's a supervisor and a subordinate," she believed, "they take the supervisor's word over the employees." She said, "They're just gonna keep lying." So, CTP was apparently exempt from outcomes of that U.S.

Office of Special Counsel finding of wrongdoing communicated in 2022 to the White House and Congress. They had absolute immunity.

Supervisor #6 was then replaced by Supervisor #7, who'd been the Team 3 Supervisor #4 already implicated by Supervisor #6 in wrongdoing. Unlike her predecessors, I never saw Supervisor #7 struggle with behaviors she'd selected that were incongruous with credible written policies and training. She showed no interest in science, our public health mission, or credible compliance. Instead, she talked about how she felt robbed for 26 years without a promotion at the FDA. She wanted more pay and she wanted a promotion and she wanted to feel appreciated and respected.

She was a Lieutenant Commander in the U.S. Public Health Service, yet, like her predecessors, continued the same inappropriate conduct. She lied and compromised our work while she proudly sat on video pointing to accolades on her uniform. Her lies often contradicted written documentation and what she, herself, had written. And what became increasingly apparent, was that those who chose to lie often didn't do it well. However, if the FDA always believed supervisors who had absolute immunity, sloppy lies wouldn't matter.

For example, on January 22nd, 2025, the same day I alerted leadership about serious indiscretions involving those 20 oral pouches products—that I was required to communicate—she filled an optional narrative in my annual performance evaluation with already disproven lies. She again misrepresented my microplastics work and again lied about what happened when I'd attempted to follow a manager's directive to ensure our work was credible. She also falsely claimed that I refused a work assignment.

She'd assigned me a review during a time I'd be on leave for over three weeks. So, I literally wouldn't be there to do the review. I'd be in Germany. She'd approved that leave that had been blocked on my calendar for months. The manager that coordinated that assignment said the review could instead go to the next person in the queue. It was the holidays and that sort of thing wasn't odd. However, two months later Supervisor #7 wrote me up for "refusing a work assignment." I didn't refuse the assignment. I asked her, based on what the manager advised in an email that she saw, and she agreed in an email to do what the manager advised. No one told me I'd be written up for asking what was advised that she agreed with. And I literally wouldn't be there.

She excluded the actual work I'd done in 2024 and dropped my performance evaluation from a 4.2/5 in 2023 to a 3/5 using that optional narrative made up of already disproven lies. Her narrative wasn't required. So, when she chose to lie again there, she targeted my career and livelihood, even if I left the agency. The union took up the case and filed a grievance. Days before I left the agency, in meetings with the union and FDA Labor and Employee relations, she haughtily backed her already disproven lies, and I've not learned of a correction to my performance evaluation. It was astonishing to watch someone in uniform lie like that when she knew we knew that she was lying. She must've known she had absolute immunity.

I'd previously reported what she did, each time, as I was required to do. So, each time what she did was allowed, that created more evidence that the agency thought that was okay. As prescribed, each time I consecutively shared evidence with our Branch Chief #7 who ignored it, then with our Division Director who ignored it, and then with our Office of Science Director #2 who allowed it. I reached out to ADR and cooperated with AHP-CREW. As the pattern persisted, CTP's Office of Management, FDA's Labor and Employee Relations, our Division Associate Director, our Branch Chief #3 and #5, and others, also participated or knew. They and our CTP Director #2, Office of Management Director #2, and the Office of the FDA Commissioner saw hundreds of pages of evidence of wrongdoing that they were required to raise until resolved but didn't. And if they wanted our work to be objective, they should've been concerned about how those behaviors affected the objectivity of our work. Every single one of them was required to raise what was happening.

However, with respect to Supervisor #7, they liked what they saw. That long search for a permanent Team 1 supervisor was over. With Supervisor #7, they found a keeper! Long before she was promoted, they saw evidence that unequivocally proved she'd lied for months in ways that affected our regulatory work, and she was rewarded. So, lying wasn't randomly applied or just tolerated, it was *integral* to doing business at CTP, and considering it was known and supported higher up, also at the FDA, at least when it came to tobacco products.

The first lie I saw Supervisor #7 tell was about that Microplastics Exposures from Tobacco Products Project. She knew better when *she also* said, "No progress has been made on this project" and put an end to those efforts. The mountain of readily available documentation made it easy to quickly dispel that apparent miscommunication. Nonetheless, for six months, she repeated that same lie, so proof that she was lying about that, as well, moved its way up the ranks.

Why had there been a turnaround with respect to that project?

Why had she said, "My goal is to help facilitate an end to this project," when just a few months earlier it was on the list of priorities for CTP's Office of Science?

31 | Oath of Office

CORRUPTION COMES IN MANY FORMS and is brought about via various sets of circumstances. We're vulnerable to corruption when the things we care about are threatened, and we can be enticed. We choose questionable behaviors to experience or obtain something pleasant or of value, or to avoid something unpleasant.

The U.S. Government is required to do things to reduce the likelihood of corruption in its workforce. Employees are vetted, for example, via background checks for circumstances or past behaviors, like addictions or gambling losses, that suggest bias or risk-taking for reward. Laws, policies, training, and external messaging indicate aboveboard behavior is required. And when entering federal service, we each sign an Appointment Affidavit that includes an Oath of Office. Each employee swears:

I will support and defend the Constitution of the United States against all enemies, foreign and domestic; that I will bear true faith and allegiance to the same; that I take this obligation freely, without any mental reservation or purpose of evasion; and that I will well and faithfully discharge the duties of the office on which I am about to enter. So help me God.

Further, 18 U.S. Code § 1001(a) states:

Except as otherwise provided in this section, whoever, in any matter within the jurisdiction of the executive, legislative, or judicial branch of the Government of the United States, knowingly and willfully—

(1) falsifies, conceals, or covers up by any trick, scheme, or device a material fact;

(2) makes any materially false, fictitious, or fraudulent statement or representation; or

(3) makes or uses any false writing or document knowing the same to contain any materially false, fictitious, or fraudulent statement or entry;

shall be fined under this title, imprisoned not more than 5 years ...

32 | More Definitions and Duties

I'D APPLIED FOR MY POSITION with the FDA during the COVID-19 pandemic in mid-2020. I'd hoped to support the agency tasked with protecting our public health. My job was to use sound and objective science to meet CTP's regulatory decision making and reporting requirements. From the beginning, that messaging was integral, but it wasn't what happened on the ground.

In the job announcement, required qualifications associated with the two levels at which the agency was hiring were listed:

GS-12: Have at least one year of specialized experience, equivalent to the GS-11 grade level in the Federal service, that includes assisting with the development of scientific experiments, conducting biological and environmental studies/investigations and following established experimental protocols.

GS-13: Have at least one year of specialized experience, equivalent to the GS-12 grade level in the Federal service, that includes investigating or analyzing scientific data from biological and/or environmental studies using new methodologies or existing techniques and preparing synopsis of data submitted for review.

That announcement listed the job responsibilities:

- *Develops plans for biology projects including estimates of personnel, equipment, materials, and schedules required to carry out plans.*
- *Establishes new criteria or extends existing methodology to adapt and modify the established guides, precedents and methods.*
- *Plans and conducts studies and investigations concerning biological and toxicological processes that affect biology resources.*
- *Develops and writes reports that include recommendations for major changes, elimination, or improvement of operations and program plans.*
- *Serves as a technical advisor on matters related to exposure and health determinations as it relates to tobacco exposure.*

A questionnaire asked applicants to rate their experience relative to things the agency determined were important for that position:

- *Assist with new processes, protocols, or guidelines related to the testing of tobacco products or other regulated products, like drugs, food, or chemicals.*
- *Review documents or studies concerned with exposure, biological, and environmental determinations of tobacco products or other regulated products, like drugs, food, or chemicals.*
- *Assist with strategies to meet short and long-range goals for the applicable programs.*
- *Evaluate biological or environmental data to determine whether data is adequate for regulatory purposes.*
- *Analyze biological or environmental investigation data utilizing new methodologies or existing techniques.*
- *Interpret and apply law, regulations, policies, standards or procedures to specific issues regarding biological or environmental studies.*
- *Search for and extract relevant information from files, documents, reports, publications, recordings, electronic databases, or other materials.*
- *Edit technical documents or scientific reports to ensure content are correct and complete.*
- *Summarize scientific review recommendations and comment on the technical deficiencies of a research project or scientific work.*
- *Prepare written reviews of biological or environmental data on the impacts of tobacco products or other regulated products, like drugs, food, or chemicals.*
- *Provide feedback on the content of technical documents that are related to the exposure, biological, or environmental effects of regulated products, like drugs, food, chemicals, and tobacco.*
- *Advise scientists or researchers on the data required to complete the review of research studies.*
- *Present at scientific meetings or within the organization on the results of scientific research.*

The interview included a timed writing exercise. I was asked, "What about CTP interested you enough to apply for this job? Where do you see your career in five years, and how do you believe working in the Office of Science at FDA/Center for Tobacco Products would advance these career goals?" I responded:

Initially, duties associated with the job description piqued my interest in the position. I am looking for a position where I can apply my combination of skills, experience, and interest in biology, toxicology, project management, and policy towards regulatory needs. I am particularly interested in being a part of analyses and decisions associated with safe use of chemicals. By this I mean that I enjoy considering and determining thresholds where effects occur. Chemicals are part of our environment, and I enjoy contributing to processes that allow us to use them safely, to avoid non-target effects.

Additionally, I find this position, in particular with CTP, interesting owing to placement of the science framed by health, social, economic, and policy aspects. Although this position does not seem to involve those aspects directly, current events associated with each area will no doubt reflect on the duties of the position. I am interested in learning more about how these aspects are informed by science and would like to help inform these issues. In five years, I hope to have gained a working fluency in the position that allows me to not only perform my duties extraordinarily well, but gain additional training and experience in these areas. I believe that CTP can help me advance these goals because I will work with environmental aspects of the tobacco industry with which I enjoy considering, in an atmosphere where exposure to work relevant to my interests in more broader aspects of the placement of that science can also be developed longer term.

They liked what they saw and extended an employment offer. They then provided a position description that further described CTP's and FDA's responsibilities:

The Center for Tobacco Products (CTP) is responsible for implementing the Family Smoking Prevention and Tobacco Control Act (Tobacco Control Act). This law gives FDA authority over tobacco products by adding a new chapter to the Food, Drug and Cosmetic Act. The Tobacco Control Act gives FDA the authority to regulate tobacco products and manufacturers based on the best available science and CTP will be responsible for both assessing and fostering that science-base. The center's actions are expected to have significant public health and consumer protection impact, and are among the most important issues faced by the agency in its long and distinguished history.

The Food and Drug Administration (FDA) is responsible for protecting the public health by assuring the safety, efficacy and security of human and veterinary drugs, biological products, medical devices, our nation's food supply, cosmetics, and products that emit radiation and for tobacco product regulation in order to decrease the tremendous toll of disease, disability and death caused by tobacco use in the United States. The FDA is also responsible for advancing the public health by helping to speed innovations that make medicines and foods more effective, safer, and more affordable; and helping the public get the accurate, science-based information they need to use medicines and foods to improve their health and to avoid or eliminate hazards associated with use of tobacco products in general and specifically among our nation's youth.

I was reclassified as a toxicologist but started as a biologist with the same duties. In that position description, my role and the percent effort I'd dedicate to the duties of my position were described:

- *The incumbent serves as a Biologist within the Division of Non-Clinical Science, Office of Science (OS) at Center for Tobacco Products (CTP), Food and Drug Administration (FDA), Department of Health and Human Services (DHHS).*
- *The incumbent will perform duties that are related to regulated tobacco products toxicity, carcinogenicity, and environmental impact.*

During about 35% of my time, I'd:

- *Plan and conduct studies and investigations concerning biological and toxicological processes that affect biology resources.*
- *Deploy data-collection systems and manipulate and analyze data from these systems.*
- *Analyze and resolve complex conventional problems.*
- *Write comprehensive reports on study findings and make recommendations.*
- *Modify and adapt standard biological techniques, processes, and procedures; assess, select, and apply precedents; and, devise strategies and plans to overcome significant resource problems related to biology.*
- *Analyze data, perform scientific tests, and conduct studies on the impact of existing or proposed management or public practices on biology resources.*

During about 35% of my time, I'd:

- *Develop plans for biology projects including estimates of personnel, equipment, materials, and schedules required to carry out plans.*
- *Inform senior biologist recommendations for modifications to project objectives and associated documentation in accordance with established directives.*
- *Conduct analyses, and review results of studies or investigations. Write reports that include recommendations for changes, elimination, or improvement of operations and program plans.*
- *Conduct project analyses and review results of studies or investigations. Evaluate the impact of the latest technological advances in biology on novel studies conducted and on current and future programs.*
- *Write reports that include recommendations for major changes, elimination, or improvement of operations and program plans.*

And during the remaining about 30% of my time, I'd:

- *Conduct project analyses and review results of studies or investigations.*
- *Evaluate the impact of the latest technological advances in biology on novel studies conducted and on current and future programs.*
- *Develop and write reports that include recommendations for major changes, elimination, or improvement of operations and program plans.*

33 | An Agency in Crisis

HUNDREDS OF SUBSTANTIVE COMMENTS over the years from scientists across the FDA point to a public health crisis. For years, the same concerns have been raised and alarms are sounded about the influence of politics and personal agendas on public health.

Since 2006, the Union of Concerned Scientists has partnered to poll federal agencies about scientific integrity. In 2022, 67% of responding FDA scientists agreed that "they can express concerns about the agency's work without fear of retaliation," up from 47% in 2006, 52% in 2018, and 54% in 2010, 2011, and 2015.

Before we celebrate 67%, we need to consider what that means.

67% agreed, but 33% of FDA scientists did not agree that they can express concerns about the agency's work without fear of retaliation. Some didn't agree or disagree. But only 67% agreed. So, consider what might've been missed, in 2022, if 33% of FDA scientists didn't express concerns about the multitudes of products the FDA regulates. The concerns of one in three FDA scientists might've been excluded from considerations of things like our food, drugs, medical devices, and tobacco products.

What about products considered in 2006, when 53% of FDA scientists might not have expressed concerns? Then, over half of FDA scientists didn't agree that they could express concerns without fear of retaliation. In 2010, 46% didn't. In 2011, 46% didn't. In 2015, 46% didn't. And in 2018, 48% didn't.

Consider the legacy when even a small percentage of concerns are excluded, over many years, from considerations of products that go on—and stay on—the U.S. market.

Also in 2022, of 239 FDA scientists who responded, 41 didn't agree that "the agency's actions were always or frequently consistent with its scientific findings." Year after year, how do agency actions inconsistent with scientific findings affect the safety and efficacy of FDA-regulated products on U.S. market?

Also in 2022, of 236 FDA scientists that responded, 50 "felt that senior leaders with conflicts of interest inappropriately influenced agency decisions." Year after year, how has each decision by each biased leader shaped the U.S. market?

Outcomes of cases heard by compromised judges are reconsidered. What about outcomes of flawed decision making associated with consumer products and our health? Is that not important?

CTP came online in 2009. However, FDA scientists alerted us beforehand about the possibilities. Only a few comments from many polls are included here, but they are representative—and they are stunning. It is stunning to see the same behaviors—known for years at the FDA—subsequently applied then allowed to continue when learned about at its new center, the Center for Tobacco Products.

For instance, in 2006, when asked to complete the statement, "The integrity of the scientific work produced by FDA could best be improved by," they said, "There is a remarkable amount of pressure placed on reviewers to find 'creative' ways to approve problematic drugs."

They said, "Twice in the past month a Division Deputy Director stated during a meeting that they were approving a drug, regardless of the medical and statistical review, without ever looking at the data."

They talked about "managers who pressure reviewers into changing their conclusions" and about "management telling staff to make changes to their reviews/letters, often repeatedly and inconsistently."

They said, "My Division/Deputy Director stated to me, 'If you don't approve this application, I am going to reassign it to another Medical Officer.'"

They said we need to, "Make [the] culture regarding saying 'no' or giving negative results more acceptable," because that is "very difficult now."

They said, "By the time the most senior review is completed, many (most) of the initial concerns/issues are obfuscated."

Again and again, year after year, they reference behaviors that counter training and broad messaging, that "management should never tell scientists what conclusions would be acceptable. Officially, the agency does not put such pressures on its scientists, and yet it has happened to me."

They said, "The annual Performance Plan should not consider supervisors opinion right & reviewer incorrect—more protection needed."

They talk about science sidelined to facilitate products moving on to the market and said, "The public should not be guinea pigs just to provide industry with revenue."

They said, "The culture at FDA is to approve drugs. To not approve a drug takes more time for the reviewer to try to gather enough information to mount a strong, defensible argument against approval."

They said, "FDA's integrity would best be improved by focusing more on ensuring product safety instead of meeting review deadlines."

They said the integrity of the science can be improved by "rewarding scientists for good science and not for arriving at the most convenient results, or for completing drug reviews quickly."

They said, "The best managers from medical review divisions to the sciences, are being reassigned to useless tasks and are being replaced with the most dishonest and unethical individuals. Those who get ahead do so by being yes-men, and by copying and pasting what the drug companies say directly into their reviews. The FDA is presently being stacked at every management level including the lowest levels based on those who will support the big companies' agenda, and the implications for safety and efficacy will be felt long into the future."

And they said, "Separating scientists from expectation that they will recommend approval" will improve scientific integrity.

Fast forward to 2022, when FDA scientists repeated concerns about compromised science at the service of industry, that "Product quality has been relaxed for approval."

They said, "Some of the decisions on approval of drugs strike me as occurring without adequate scientific backing for the decision."

They said, "Science is now basically a show game of testing procedures/policy, with little real study of safety."

They said, "We are told to work to help industry approval and we are not agents helping do the peoples business."

They said, "There has been a tremendous job exodus from my center of subject matter experts who ask questions about products, as they may appear to retard the approval speed."

They said, "Regarding scientific integrity, we are no longer able to justify our work in terms of the 'public health,' but mainly now on promoting assistance to industry and in reviewing faster and faster."

They talk again about coercive leadership behavior. They said, "I was told to stop work on one paper due to concern that the results would not be consistent with 'messaging.'"

They said, "Bullying is extensive at my center. Unfortunately our Commissioner stated: 'If you find bullying report it to your supervisor,' but often your supervisor is the bully."

They said, "Agency integrity is determined more about upper management keeping a lid on everything."

And they expressed concerns about the lack of protections for those who raised concerns. They said, "Although there are whistleblower policies that are supposed to protect employees, it has been my experience that they are not often enforced adequately and they do not shield the whistleblower from undue harm."

They said, "Regarding whistleblowers, although protections for them have improved to some extent in recent years, those people almost always suffer great hardships."

They said, "I have made two whistleblower complaints in regard to the manipulation of personnel, the second to the highest authority in our office. One complaint was in regard to fraudulent scientific work that was investigated and proven to be relevant. No action was taken to review whether or not that work was more than just a current issue, even though complaints have been occurring for many years."

And they talked about litigation and static science. They said, "Regulations and sometimes litigation risks can impact scientific integrity. Regulatory changes cannot keep pace with the progress of science so sometimes what's best from a scientific standpoint is not feasible given the current regulations. The agency is also being given less deference in courts so risk of litigation and potential impacts of losing a court case can hinder scientific decisions."

They said, "I'm concerned that senior management may be influenced by the threat of litigation, out of the concern that such litigation might ultimately lead to limits on the agency's regulatory authority by the court."

Good work is no doubt being done by well-intended employees and good scientists at the FDA. It is critical to our health, wellbeing, and economy that we parse and protect those efforts from bad actors. I experienced good work outside of CTP in FDA-wide efforts related to nanomaterials and microplastics where interactions were professional, respectful, and aligned with the agency's public health mission.

I experienced pockets of that in CTP, for example, with our Office of Health Communication and Education and with respect to some policy considerations in the Office of the Center Director. Some good science even happened in my division and branch, where wrongdoing was pervasive. Outcomes of that research just didn't objectively inform our regulatory work, so our tobacco regulatory work was immune from even our own science. Research in my branch was redirected to questions less likely to counter findings that support industry.

I contacted the U.S. Office of Special Counsel on August 18th, 2024, about behaviors that hadn't been corrected since their December 15th, 2022, finding of wrongdoing. I also provided evidence that politics influenced our work. Certain regulatory decisions were fast-tracked in response to Congressional inquiries then delayed until after the 2024 presidential election.

Start, stop, start, stop, and those delays weren't inconsequential. Delayed regulatory health decisions deprived the public of protections and important disclosures. Fast-tracked work was sloppy and intentionally excluded considerations that would be difficult—and therefore time-consuming—to properly address. That also risks people's health, especially when that became the precedent we were expected to repeat.

I also alerted Congress. Senator Amy Klobuchar was on the U.S. Senate Committee on the Judiciary that heard that June 12th, 2024, testimony from our CTP Director #2. That committee was paying attention and their questions hinted that they knew at least some of what was going on. And I was lucky to live in her state. I'd read some lawmakers would only respond to their constituents.

I contacted Senator Klobuchar on August 19th, 2024, and she was responsive. I then mailed her several hundred pages of organized evidence signed for at her Minneapolis office by "A.K." On September 6th, 2024, she asked me to meet with her Senior Counsel on the U.S. Senate Committee on the Judiciary. I emailed him on September 8th, 2024, and the next day he replied, "I am looking forward to finding a time to discuss." However, we never talked. We'd exchanged emails but hadn't found a time to meet, and I didn't want to push. I was nervous.

Every now and then, I remembered that tobacco is a high-dollar commodity. Well over a hundred million dollars a year went to most states just from tobacco taxes. Tobacco companies lobby Congress and tobacco money supports Republicans and Democrats. I was just a measly little scientist trying to abide by my silly oath, written policies, and training that aligned with our public health mission. There is a lot of energy working against that.

And the more I interacted, the more I saw that looked sketchy. Some meetings and interactions, that probably shouldn't, seemed inappropriately corporate. For instance, the affiliation on my name tag and proximity to our CTP Director #2 during that American Public Health Association Meeting had people handing me business cards and wanting to talk. Some of that interest was aboveboard, but some of it wasn't. And when I asked about following up with contacts who I could tell did credible work, like researchers at universities with very public research records, I wasn't allowed to interact. Some of those researchers could've helped inform CTP's regulatory considerations of specific products we were considering at that very time.

On October 3rd, 2024, the U.S. Office of Special Counsel responded, "It appears that CTP is already in the process of developing agency policies that will provide for more robust consideration of the cumulative effect of all possible environmental impacts of granting tobacco marketing orders." That was great, and although cumulative effects are important, the issues and context were much bigger. CTP continually told Congress and the public that our science was already robust, and it wasn't with respect to a lot more than cumulative effects.

The U.S. Office of Special Counsel acknowledged the concern that CTP would continue to endanger the public health and safety while continuing without written processes and policies, and cited CTP's self-imposed July 1st, 2025, deadline as appropriate to implement new policies that would rectify concerns. They'd *again* allow CTP to self-correct. They acknowledged, "CTP continues to issue tobacco marketing orders in the meantime."

In hindsight it seems naïve, but I believed I'd done my due diligence. I thought enough people knew, that the record would be corrected, and that the credibility of our work would improve. So, I told Senator Klobuchar and my leadership that I'd wait on those July 1st, 2025, improvements. My expectation was that we'd do no additional harm in the meantime, for example, by granting marketing orders to more products via work that didn't align with FDA's Scientific Integrity Policy.

Implementing that policy would instantly improve the credibility of our work and therefore immediately reduce new risks to the public. Explicitly advertised to get us back on track, CTP also had a new Five-Year Strategic Plan with five goals:

1. *Develop, advance, and communicate comprehensive and impactful tobacco regulations and guidance*
2. *Ensure timely, clear, and consistent product application review*

3. *Strengthen compliance of regulated industry utilizing all available tools, including robust enforcement actions*
4. *Enhance knowledge and understanding of the risks associated with tobacco product use*
5. *Advance operational excellence*

That was good news! That strategic plan came out on December 18th, 2023, and incorporated recommendations from the Reagan-Udall Foundation report. So, by October of 2024 some improvements might've been apparent. But I didn't think things would get worse. I believed there was some vein of accountability somewhere that held us to incessant messaging about our high standards.

For example, our CTP Director #2 had just talked about that strategic plan on September 25th, 2024, at the Global Tobacco and Nicotine Forum 2024, in Athens, Greece. He touted pillars of scientific integrity in that plan, pinpointing the importance of robust science and transparency. That plan even referenced risks associated with the *use* of tobacco products (that some leadership falsely claimed we didn't consider). That reinforced the need for disclosures about secondhand and thirdhand emissions that included lead and other harmful constituents, as well as our need to know what those oral pouch products were made of.

That said, he'd issued his inaccurate SDR determination on December 22nd, 2023, just four days after the Reagan-Udall Foundation report came out and seven days after the U.S. Office of Special Counsel's disclosure of wrongdoing at CTP.

During that September 25th, 2024, meeting, our CTP Director #2 said, "And I will say what I've said again, that we're not going to jeopardize the science, and we also need to make sure that our actions are legally defensible, and that takes time."

He said, "I didn't write the law. I am in charge of implementing it. And so that is what we are going to do. I don't look good in orange, folks. So, we're gonna make sure we follow the law."

He talked about "rigorous scientific evidence" and "being responsible to make sure we're maximizing impact of these products for public health while minimizing the adverse effect, particularly among youth."

He said, "We want to make sure that we are abiding by the law and what Congress intended, and so we have a very clear scientific bar."

He said, "We're not going to jeopardize that." He said we're not going to jeopardize that "high *necessary* scientific bar."

Also in hindsight, watching interactions at that meeting has me wondering. Our Office of Science Director #1, who'd left CTP in July of 2022 to work for a tobacco company, was at that meeting. He'd been our Office of Science Director #1 since 2017. He said to our CTP Director #2, "I was a little surprised to see you not mention *NYTS*."

He was likely referring to data from the *National Youth Tobacco Survey*. Earlier that month, on September 5th, 2024, the FDA and CDC released data from the *National Youth Tobacco Survey* about the use of products like e-cigarettes and those oral pouches sold but not yet approved for sale by CTP. Also on September 5th, 2024, an article in *The New York Times* said, "The percentage of middle and high school students reporting that they vaped tobacco products declined to about a third of the peak levels in 2019, a new survey shows."

Our former Office of Science Director #1, continued, "You guys just released really good data showing that youth vaping numbers have dropped yet again, looked, also released data on nicotine pouches which show that their use amongst youth is still very, very low. And again, not seeing that in the presentation [our CTP Director #2 had just delivered], but also just thinking about the broader sort of communications, you, they come out of CTP. I mean, obviously in 2018, the term *youth vaping epidemic* was generated from the agency to sort of show the uptick. But I wonder, now that we're seeing this sharp decline, why there's not more communication about that, and, you know, sort of, again, a sort of messaging also to adult users that these products can be innovative products that help them get a safer form of nicotine."

The company that applied for marketing orders for those 20 oral pouch products was a subsidiary of the company he currently worked for. Our reviews of those applications were ongoing during that September 25th, 2024, meeting. And it seemed like the discussion in the room was framed by the expectation that those products would receive marketing orders, not that they were currently undergoing a robust or extensive and objective scientific review.

Our CTP Director #2 replied, "Yeah, so I certainly appreciate the comment, and if you've got suggestions of better ways to amplify it, I'm all ears. We did a press a release, the National Press Briefing, and I did a national media tour. I sat for five hours talking to every acronym and media across the country. And so we definitely do our best to help promote those findings and I ultimately think it is a good thing. We're 70% lower than we were at the peak. I do think we can go lower. We certainly don't want to rest on our laurels. We see where we are with other products, that are lower than the current estimate for e-cigarettes. But bottom line is, I've been very vocal that I think it's a monumental public health win. How did that get accomplished? It was coordinated efforts. There was no single panacea there either, but I'm hopeful that it will continue to decline, but to do that we got to put the pedal to metal on the things we know work."

What percentage of youth still using addictive products with uncertain health outcomes constitutes a "monumental public health win," and should our CTP Director #2 be "all ears" with respect to what industry wants while sidelining harms communicated to him by his own scientific staff? Maybe consider that reduction, in an exploding ENDS market that our public health agencies ignored, a *reduction* instead of a "monumental public health win," especially when youth are still addicted and newly becoming addicted. And was it a monumental public health win that youth used some products less than e-cigarettes? These were weird comparisons.

Our CTP Director #2 continued, and for some reason mentioned our review of applications. He said, "In terms of changing the calculus of application reviews, you know, youth use is still a concern, and so I wouldn't expect some major shift in any of our decision making. Youth use is still too high, and we got to make sure that we don't see any uptick, so we're playing whack-a-mole. But bottom line is I think it's good news."

Why had he mentioned "changing the calculus of application reviews"? Is that what he thought was expected? That guy from that tobacco company who used to be our Office of Science Director #1 only mentioned messaging, and shortly afterwards, the calculus of our application reviews had, indeed, changed so we'd grant marketing orders for those additional six *flavored* oral pouch products appealing to *youth*.

Our CTP Director #2 continued, "On the nicotine pouch front, I think it's good news as well."

What about that was good news? How could the uptick in their sales, or *any* sales or *any* use by youth or by *anyone*, be "good news" when those products hadn't even passed our rigorous—or any—scientific review? None of those products had even received marketing orders!

He continued talking specifically about oral pouch use, "But it's also slowly upticking, and so I'll just remind folks that let's not revisit the past. And so, I think there was some key lessons learned on the e-cigarette front, particularly with regard to youth access and use. And I think that we would be remiss if we didn't consider that on the issue of nicotine pouches. There's certainly potential to help adult smokers quit, but again, we're weighing benefits and risks. And so you got to do everything you can to mitigate that youth use, and so I'm very hopeful, given the recent lack of change over the past year. But I'm also mindful that we've seen a steady uptick over the past several years. And so what we don't want to do is to see that rise further. And so I think the onus is on all of us, including many of the people in this room, to make sure that we ride that success, continue to see declines in youth use while maximizing opportunity for adult smokers. And for our part, we're committed to verbalizing that in whatever way is possible. And if folks have suggestions around comms, again, I'm all ears, but I'm traveling three, four speaking engagements a week and doing everything we can on press, including a release every week. There's only so much we can do, but definitely open to opportunities for further engagement to maximize our comms impact."

Yes, let's not revisit the past. He knew sales were more than "upticking." However, his CTP sausage-makers said we needed to use older data. Why?

When product reviews hadn't been finalized, we were advised to say, "We're not able to comment on products currently under review." However, this time, our CTP Director #2 said to the applicant that news about those products was good, even though none had received marketing orders and we hadn't even completed our scientific reviews and assessments of those products.

Another audience member then asked, "You know, it's taking a while. I think the pouches have been submitted about four years ago. Is there a reason other than not having enough money and reviewers to get them sort of authorized?"

Our CTP Director #2 definitively replied, "Nope. That's it. We'll get to it. But we don't have the resources."

"We'll get to" getting "them sort of *authorized*"!? Wrong answer. Shouldn't that've been, "We'll get to doing robust scientific reviews and assessments"?

The audience member then said, "I'm going to send you a check. What do you need?"

"So, $114 million," our CTP Director #2 said, "is what we've requested from Congress."

The audience member replied, "Oh, oh, we'll get that from someone else then," and said, "[our Office of Science Director #1 who'd left to work for that tobacco company] will send you that, [him], over there."

People laughed and our CTP Director #2 responded, "Yes, yes, yes. We may have to downgrade the next conference. Everyone can chip in. We can be at the Best Western in Montana, and we'll use the added resources for CTP."

The microphone was then passed to another audience member who said, "We have developed a fully biodegradable plant-based water-dispersed, dispersing substrate with aim to replace cigarette filter, which is based on cellulose acetate, which is a single-use plastic, which is the number one most littered item in the world." He asked, "Is CTP currently considering a biodegradable solution to replace a number one most littered item in the world that impacts U.S.A., and if not, why not?"

"No, it's not currently a priority for us," our CTP Director #2 uncharacteristically coldly replied and said, "I'll say that we've got multiple things and limited staff and finite resources and that currently is not a priority for us."

"Okay, well, [CTP Director #2]," the moderator then said, "Thank you as always. I'm glad we didn't have you in the Best Western today. We can arrange that if you want. We can get you over there."

Someone off mic apparently offered our CTP Director #2 an article. He said, "I don't need *The New York Times* article, thank you," and stepped off stage.

I wonder if they were just as excited about communicating what has been learned since. For example, on July 9th, 2025, *The New York Times* article, "Just How Harmful Is Vaping? More Evidence Is Emerging," said, "A new study that found high levels of heavy metals in popular nicotine vapes adds to concerns about the products." Where is CTP's and the industry's communications piece about that?

Congress repeatedly expressed frustration about delays in marketing orders for ENDS and oral pouch products. On September 10th, 2024, they'd just asked our CTP Director #2, "Why hasn't FDA issued a marketing order for a single synthetic nicotine product?" And he'd asked them for $114 million.

Then, on September 25th, 2024, our former Office of Science Director #1 stood in front of him and also asked about oral pouch products, currently under review, that were produced by a subsidiary of his current employer. Even if he wasn't pressuring our CTP Director #2, and I'm absolutely not saying that he was, he knew where the bodies were buried. He knew particulars that would fairly undermine CTP's regulatory authority. He wouldn't need to say anything about that—our CTP Director #2 knew that he knew.

Did our CTP Director #2 walk away from all these sorts of interactions feeling that he needed to placate, in this case Congress and former CTP employees now working for the tobacco industry, by granting marketing orders for those 20 oral pouch products?

At that September 25th, 2024, meeting, he'd already made up his mind about those products. He didn't talk about those products like he talked about our other work. CTP had decided to grant those marketing orders before we'd been able to complete our rigorous scientific reviews and assessments.

On October 3rd, 2024, the U.S. Office of Special Council wrote, "It appears that CTP has broad discretion to make determinations on tobacco marketing applications and you are not in a position to be able to evaluate the factors that contributed to the decision to delay certain determinations until after the 2024 Presidential election."

I'd not asked "to evaluate the factors that contributed to the decision to delay certain determinations until after the 2024 Presidential election."

My concern was that it was happening. My training indicated it should not.

On April 20th, 2022, the U.S. Government Accountability Office had published a report, "Scientific Integrity: HHS Agencies Need to Develop Procedures and Train Staff on Reporting and Addressing Political Interference."

They said, "Since 2007, Congress and multiple administrations have taken actions to help ensure that federal science agencies have scientific integrity policies and procedures in place that, among other things, protect against the suppression or alteration of scientific findings for political purposes." They found that the FDA didn't, even after attention from Congress and multiple administrations since 2007. They said that the FDA "did not identify any formally reported internal allegations of potential political interference in scientific decision making from 2010 through 2021," essentially because there was no process to do that. The report states that "the absence of specific procedures may explain" that.

You think?

So, what actions had Congress and multiple administrations taken?

That report defined scientific integrity as "the use of scientific evidence and data to make policy decisions that are based on established scientific methods and processes, are not inappropriately influenced by political considerations, and are shared with the public when appropriate." And as an outcome of that 2022 report, the FDA implemented a new Scientific Integrity Policy in December of 2023 that we were all required to follow. That policy didn't refer to only political interference. Among other things, it reiterated other tenets of scientific integrity well-known to any trained scientist or logical person, such as accuracy, objectivity, transparency, and protection against any inappropriate influence. However, CTP still sidelined our public health mission for politics.

Results of the 2024 Federal Employee Viewpoint Survey raised more red flags in CTP's Division of NonClinical Science where I worked. The survey was administered from May 20th, 2024, to at least July 5th, 2024. I believe the closing date might've been extended a bit. Seventy-three of us, or 75.3% of the division, contributed to those survey results. CTP's Office of Science had 399 contribute and a 69.9% response rate. CTP had 765 contribute and a 69.2% response rate. And the FDA had 13,390 contribute and a 69.6% response rate.

The survey is meant to identify areas in need of improvement.

During that survey window, on June 12th, 2024, our CTP Director #2 had testified in front of Congress. I don't recall being alerted about the hearing beforehand or afterwards, and I'm certain I would've tuned in. Our annual Office of Science picnic was scheduled for the same day, so many of us were distracted and unaware. I learned about that hearing later, by happenstance, via an online news search for "FDA tobacco" I regularly did to keep informed.

I remember, at first, feeling betrayed that no one alerted us about that hearing. This was big and we should've known. Then I watched the recording of that hearing. Despite the audacity of everything that'd come before, I remember the shock of watching our CTP Director #2 lie to Congress.

He said, "We have a premarket paradigm to review the scientific evidence, to understand what is in the product itself and to make a determination based on the science, whether they meet the appropriate statute by law."

He said, "As a scientist myself, I'm committed to following the science, and when I took this job, I vowed that and I'm continuing it now," and then confirmed, "We have an important paradigm in place."

He said, our "scientific review, to make sure that we're doing a robust review of the evidence," includes "the toxicologic risks and various other risks at the population as a whole."

"The most commonly used product among kids" is "where we prioritize the resources," he said, "So, it's important that we follow the science."

I'm certain that his testimony influenced our survey results. He knew that we knew that he knew that didn't happen in our division. However, some staff submitted their survey responses unaware of that testimony and those lies. If we'd been made aware, those survey results might've been even worse.

In two cases, responses in our division were more positive than the Office of Science, the report said, because results differed by three or more percentage points. 88.9% of our division said, "I identify with the mission of my organization," compared with 83.8% in the office. And 93.0% of the division said,

"Employees in my work unit support my need to balance my work and personal responsibilities," compared with 89.8% of the office.

And I agree!

Without a doubt, the most consistently expedient and positive responses from leadership were to requests to take leave. That's the one thing no individual in leadership lagged on granting. And in that context, they were always respectful and professional, so we knew they had it in 'em! And we were awarded a lot of extra award leave and early dismissal. It was almost like they didn't want us there.

And many of us did identify with the mission of our organization—that's what I signed on to do. The misalignment of those of us aiming to do what we signed on for with those who didn't is what led to problems.

So, many of those other survey responses, even though they were weighted by those who were complicit, looked very bad.

Compared with 75.4% of FDA employees, 58.1% of our division said, "Employees are recognized for providing high quality products and services." So, outcomes of compromised leadership were apparent. Compared with 73.3% of FDA employees, 58.2% of our division said, "My organization's senior leaders maintain high standards of honesty and integrity."

How would our health benefit if we'd, instead, provided high quality products and services?

Compared with 74.4% of FDA employees, 50.6% of our division said, "I feel encouraged to come up with new and better ways of doing things." So, our habit of repeating past mistakes and indiscretions was apparent. Compared with 78.1% of FDA employees, 69.2% of our division said, "Employees in my work unit incorporate new ideas into their work." And compared with 70.9% of FDA employees, 51.4% of our division said, "Management encourages innovation."

How would our health benefit from management, instead, encouraging—or requiring—the best, most up-to-date science?

Compared with 63.7% of FDA employees, 39.1% of our division said, "I can make decisions about my work without getting permission first." So, tactics used to force the current thinking on scientists hired to do objective science were apparent. Compared with 76.3% of FDA employees, 53.7% of our division said, "I have the autonomy to decide how I do my job." Compared with 74.3% of FDA employees, 58.6% of our division said, "I can influence decisions in my work unit." Compared with 81.7% of FDA employees, 66.8% of our division said, "I can be successful in my organization being myself." And compared with 71.7% of FDA employees, 53.5% of our division said, "My talents are used well in the workplace."

How would our health benefit if scientists hired to be scientists were allowed to work as scientists?

Compared with 88.4% of FDA employees, 69.2% of our division said, "My organization is successful at accomplishing its mission."

Consider what's been missed when so many regulatory scientists can't do their jobs?

Compared with 67.0% of FDA employees, 53.2% of our division said, "In my organization, arbitrary action, personal favoritism, and/or political coercion are not tolerated." So, biases and the status quo were apparent. Compared with 66.8% of FDA employees, 53.8% of our division said, "Management makes effective changes to address challenges facing our organization."

How would our health benefit if our work had, instead, been objective?

Compared with 69.1% of FDA employees, 48.0% of our division said, "Information is openly shared in my organization." So, efforts to silo employees by complicit leadership were apparent. Compared with 72.3% of FDA employees, 55.9% of our division said, "Managers promote communication among different work units (for example, about projects, goals, needed resources)." Compared with 64.5% of FDA employees, 45.6% of our division responded positively to, "How satisfied are you with the information you receive from management on what's going on in your organization?" Compared with 78.2% of FDA employees, 66.7% of our division said, "Managers communicate the goals of the organization."

How will our health benefit when transparency and accountability preclude biased leadership from enforcing the status quo?

This raised concerns in the Office of the FDA Commissioner and prompted intervention by the FDA's Office of Planning, Evaluation, and Risk Management (OPERM).

We already knew about issues across the FDA. So, imagine how alarming it was that CTP lagged behind the FDA, that our Office of Science lagged behind CTP, and that our division, where many of CTP's scientific, tobacco regulatory determinations happened, lagged behind the Office of Science. We were in the bottom of the very deep FDA barrel.

But this should've come as no surprise. Just two years prior, the U.S. Office of Special Counsel disclosed wrongdoing and retaliation against scientists that raised concerns about scientific integrity in our division. We were regularly queried for input about our regulatory tobacco work, for instance, as part of the Reagan-Udall Foundation's "Operational Evaluation of FDA's Tobacco

Program" and from CTP's Office of Science, with constant promises that known problems would be corrected, but they weren't. We received training about where to report if issues came up again, then our CTP Director #2 lied as part of that Scientific Dispute Resolution process they pointed us to. So, they carried on the same way while they asked us for input.

In some cases, there was no option to contribute anonymously, and that put staff at odds with management. We'd be honest about questionable practices. Then, not only would those questionable practices continue, but so would new questionable practices that obfuscated those questionable practices, like newly limited communications, fewer meetings, and notably less informed work. Over time, it seemed more like management sourced that information to learn about issues they needed to make better efforts to hide.

And even purportedly anonymous input wasn't. For instance, when an information-gathering session advertised as "anonymous" began, photos and firstname.lastname@fda.hhs.gov email addresses of the eight anonymous participants were displayed on the screen. Regardless, the analyst running the call steamed ahead. We broke in to say something advertised as anonymous wasn't. The analyst said we could all hang up, and she could restart the call. She said that would fix it.

Really?

We all kind of chuckled. We'd and she'd already seen the faces and names of all participants. That "anonymous" information-gathering session about our tobacco regulatory work was, therefore, pretty quiet.

Supervisor #2 said management's focus was on *completing* things like these surveys, not on learning from what was shared or making improvements. He said their performance was based on reportable metrics, like what percentage of the employees they oversaw participated in surveys like this, completed mandatory trainings, and signed timely tobacco product reviews and personnel documents. The content didn't matter. For example, we didn't need to attest to the accuracy of our tobacco product reviews or receive accurate performance evaluations. We just needed to sign to complete the process.

Metrics were what Congress used to evaluate CTP's success. For instance, Congress looked at how many tobacco product applications had been processed and how many products, particularly ENDS, had been granted marketing orders. When those numbers weren't what they wanted, we could provide other metrics. CTP could report numbers of new staff hired and that participated in surveys purportedly aimed at improvements. CTP counted communication and education outcomes and talked about those in detail. However, CTP didn't talk about our science, about written guidances we hadn't produced because they'd show we didn't validate data or apply recognized thresholds in unscientific determinations involving a fraction of constituents CTP was supposed to consider.

Anyway, we were offered incentives to complete those surveys. If participation surpassed a certain percentage, for example, we'd get paid time off. Of course, over time as staff grow weary of incessant reporting alongside deteriorating work conditions, they'd stop responding. Our Office of Science Director #2 was once so desperate for a high participation rate that he sent an email with the subject "Tornado" as a swath of deadly storms passed through areas where some staff lived. I quickly opened that email, where'd he'd written, "Now that I have your attention," complete this survey.

On December 3rd, 2024, our Division Director wrote, "The [Federal] Employee Viewpoint Survey report indicates that there are several areas of in which we can improve as a division. We take these results very seriously."

Employee concerns were chronically buried or unaddressed, and this time, compared with 92.7% of FDA employees, 72.0% of our division said, "My organization shares results (for example, town halls, email, distribution of reports) from the Federal Employee Viewpoint Survey (FEVS)."

He continued, "As a follow up to this [Federal] Employee Viewpoint Survey, we are working with OPERM to further survey our division. We would like to use the results of this additional survey to get specific feedback from Division of NonClinical Science staff and to gain insight into issues and concerns division staff may have. We will use this more detailed survey to formulate key action items for our leadership staff and our division as a whole."

They'd already asked so many times, and they'd already been told so many times. They already knew.

He continued, "Please look out for a survey link coming from OPERM in the coming days. I would be grateful for your feedback. A high response rate is very important so that we can get a better understanding of Division of NonClinical Science staff concerns and other feedback."

Despite concluding, "Thank you all for everything that you do," he continued supporting all that bad behavior.

On August 15th, 2025, the Office of Personnel Management announced that it had cancelled the 2025 Federal Employee Viewpoint Survey.

34 | American Guinea Pigs

THINGS WERE QUIET WHEN I RETURNED from leave on January 13th, 2025. Many took leave that time of year, so that wasn't odd. I caught up on three weeks of emails, did some training, and jumped back into my work.

Then, at 11:42 am on January 16th, 2025, our Office of Science Director #2 sent an email to everyone in the office. The subject line said, "CTP Issues MGOs to [tobacco company] for 20 [product name] Oral Nicotine Pouch Products."

What?

How?

We hadn't finalized the review or completed the assessment!

Nonetheless, it said, "Today, FDA issued marketing granted orders (MGOs) to the [tobacco company] for 20 [product name] oral nicotine pouch products." Holy smokes the marketing orders had been granted, and I was named in that email that said that determination had been made by CTP scientists "following an extensive scientific review."

But that wasn't true.

That email said, "This is the first time FDA is authorizing a nicotine pouch product. The 20 products authorized today can now be legally marketed in the US."

Delays in CTP's regulatory review process meant Americans used products for years before the agency made decisions about marketing. For years, those products had already been sold in the U.S. If the FDA had decided not to authorize the sale of those 20 products—products popular with youth—it would be fair to hold the agency responsible for any concerns it cited in its decision to deny marketing orders. And that argument stands for many decisions the FDA has made to grant delayed marketing orders after products were already sold in the U.S.

In this case, if the FDA hoped to avoid litigation, it would need to issue marketing orders. How could they say no?

This further calls into the question the FDA's post-marketing regulatory role, where the agency polices products following its own determinations that they can be marketed. Would an agency trying to protect itself from litigation, from losing credibility, and from losing its mandate, objectively monitor outcomes of products it put on the market? Would involved individuals or anyone be more supported for coming forward or for burying problems?

We know a lot about combustible tobacco products. They all expose users and non-users who share those spaces, during and after those products are used, to a long list of harmful substances. Lead, arsenic, polycyclic aromatic hydrocarbons, particulate pollution, and formaldehyde are among those. In addition to the harmful products of combustion, the tobacco curing process can increase cancer-causing chemicals called nitrosamines. Those tobacco-specific nitrosamines, with the harmful compounds emitted via combustion, are credited with the debilitating and deadly effects of combustible tobacco products.

In this case with those 20 oral pouch products, and with ENDS, unfortunately, pressures to move those products on the market also compromised the agency's objectivity. CTP leadership said our goal was to move users *down* the risk continuum, away from combustible products. The key is that the continuum is made up of products we understand well enough to make those comparisons. Otherwise, we move users *up* the continuum with respect to uncertainty related to other potential risks and when uncertainties aren't addressed those users become—addicted—guinea pigs.

This was bad. So, I reached out to our Office of Science Director #2 and his scheduler set up a meeting for 10:15 am on January 21st, 2025.

Also on January 16th, 2025, the FDA published a press release, "FDA Authorizes Marketing of 20 ZYN Nicotine Pouch Products after Extensive Scientific Review—Agency Will Closely Monitor Youth Use and Company's Compliance with Marketing Restrictions."

The first paragraph of that press release said, "Today, the U.S. Food and Drug Administration authorized the marketing of 20 ZYN nicotine pouch products through the premarket tobacco product application (PMTA) pathway following an extensive scientific review. This is the first time the agency has authorized products commonly referred to as nicotine pouches, which are small synthetic fiber pouches containing nicotine designed to be placed between a person's gum and lip."

"Extensive scientific review"?

Oh no.

Extensive is defined as "having wide or considerable extent," with synonyms like *comprehensive*, *broad*, and *sweeping*.

Scientific is defined as "of, relating to, or exhibiting the methods or principles of science," "conducted in the manner of science or according to results of investigation by science," or "practicing or using thorough or systematic methods."

The definition of *science* is "knowledge or a system of knowledge covering general truths or the operation of general laws especially as obtained and tested through [the] scientific method."

And the *scientific method* is defined as "principles and procedures for the systematic pursuit of knowledge involving the recognition and formulation of a problem, the collection of data through observation and experiment, and the formulation and testing of hypotheses."

A *hypothesis* is "a proposed explanation for something (such as a phenomenon of unknown cause) that is tentatively assumed in order to test whether it agrees with facts that are known or can be determined."

What kind of "synthetic fiber"?

We'd not even discussed that level of specificity. We never even confirmed the fibers were synthetic. If we'd hoped to apply logic and empirical data to determine any sort of outcome associated with those products, one fundamental part of our sweeping, broad, or comprehensive reviews would need to be knowledge about what the products were made of.

The second paragraph of that press release said, "The FDA determined that the specific products receiving marketing authorization met the public health standard legally required by the 2009 Family Smoking Prevention and Tobacco Control Act. This standard considers the risks and benefits of products to the population as a whole."

How could such a determination be made without considering what those pouches were made of?

The third paragraph said, "Among several key considerations, the agency's evaluation showed that, due to substantially lower amounts of harmful constituents than cigarettes and most smokeless tobacco products, such as moist snuff and snus, the authorized products pose lower risk of cancer and other serious health conditions than such products. The applicant also provided evidence from a study showing that a substantial proportion of adults who use cigarette and/or smokeless tobacco products completely switched to the newly authorized nicotine pouch products."

Constituents?

Again, how could such a determination be made without considering what the pouches are made of?

That press release ended with, "Today's actions are the latest of many the FDA has taken to ensure all new tobacco products marketed in the U.S. undergo science-based review."

No. No. No. No. No.

Science-based?

Ensured? After having been sold for years?

Bad FDA!

This was weird. Normally, to keep us all looped in, there would've been email communications as work products moved forward in the review process. I'd just gone through my emails and didn't recall seeing anything.

I looked again, and there was nothing. How could there be nothing?

Then, at the end of the day, at 4:54 pm, on Friday, January 17th, 2025, I received another email. Our Office of Science Director #2 emailed those of us involved with the regulatory clearance of those 20 products. He said, "Hello all, Just want to send you all a note of thanks for all your efforts to get the [oral pouch product name] MGOs [Marketing Granted Orders] over the finish line. It certainly wasn't easy and required a lot of last-minute efforts but many of you to get it out this week. So, thank you and I hope you enjoy the long weekend."

Then at 5:33 pm, our CTP Director #2 sent me an email with the subject line "Thank You" and wrote, "Hi Christy, I'd like to extend a personal note of gratitude to you for all your work as part of the review team on the recent [oral pouch product name] authorization. This agency action would not have been possible without all your time, expertise, and unwavering dedication. I hope you have a stellar long weekend!"

This was all weird.

This needed to be documented in writing. So, I postponed my January 21st, 2025, meeting with our Office of Science Director #2. Instead, at 10:03 am on January 22nd, 2025, I emailed him and our CTP Director #2. I told them the assessment hadn't been done properly. I mentioned risks to the public that we were obligated to protect and inform, and I summarized what happened. Some of that wasn't new to them.

"I need to be sure I make every effort to do my due diligence," I said. "Please frame the events below understanding that I must follow our policies, training, and directives in support of our public health mission."

I said, "We had no comparator product, the market share was increasing, and we did not know the materials that constituted the pouch."

I said, "I was repeatedly told by my Secondary Reviewer #3 that the pouch completely dissolved and therefore there would be no issues with use or disposal of the pouch." I mentioned my trip to the Mall of America, after I'd left the American Public Health Association meeting, to debunk those claims. Our CTP Director #2 was at that meeting too, so he'd seen research about those pouches and would know we'd need to pause those marketing orders.

I said, "No reviewer had apparently learned this [that the pouch didn't dissolve] during the cycle one review and during the cycle two review before I came on board. I joined the review team halfway through cycle two and this sort of issue is common, something is missed during a previous review that we are not allowed to raise when we discover it. This violates our policies and mission and does not reflect our training. We could have simply asked the applicant for information about the pouch materials. We could have discussed that information in general terms in the assessment and listed details about those materials in a confidential appendix, as we'd done before, respecting confidential business information."

However, evidence indicates people did know.

I said with respect to the assessment, "In a public document with my name, credentials, and experience listed, and as our microplastics subject matter expert who leads our center's microplastics research efforts, I was not allowed to include mention of the pouch materials, even know what the pouch materials constituted, or include any mention of substantiated issues associated with oral pouch products. Information I tried to include was removed."

I said, "I know from experience that we are not using robust or comprehensive scientific approaches. As I've tried to do that as per our policies, training, and directives, I've experienced an ongoing barrage of retaliation."

I said, "I know after working for several years on a project that I initiated due to concerns with consumption of plastic during oral pouch use, that users would also likely be consuming plastic during use. Masticating the product in the oral cavity [or simply holding it there] creates an environment that favors polymer breakdown and consumption (e.g., with temperatures and enzymatic activity that increase pouch degradation)."

Our CTP Director #2 was familiar with that work. I'd increasingly reviewed pending agreements, with respect to microplastics exposures, with his staff in the Office of the Center Director. And I was sensitive to representing uncertainties associated with the effects of microplastics exposures. I said, "Even if the body of literature was inconclusive about such matters, some mention should be made alongside that uncertainty." Regardless, an extensive scientific review required we knew what people were putting in their mouths.

I asked what they thought we could do.

Despite the seriousness of what he'd learned, our Office of Science Director #2 didn't respond or cooperate at all. Our CTP Director #2 was also uncooperative, despite his thanks for my expertise and unwavering dedication just a few days before. I shared evidence of serious wrongdoing with them both. But like others before him, our CTP Director #2 directed serious regulatory science issues to a group not designed to handle those issues. He chose not to engage about those issues.

Instead, in an email he sent at 4:18 pm on January 24th, 2025, he wrote, "I have forwarded your emails to the FDA Anti-Harassment Program for review. I want to assure you that we will implement the decisions made by the Anti-Harassment Program as appropriate."

The ridiculousness of what he wrote made me chuckle, but this wasn't funny. It was disgraceful.

I'd tried to do credible work for four years under corrupt leadership. I'd engaged directly with them—I'd cooperated—for over 21 months via internal processes that proved performative. It would be ridiculous for me to spin wheels with them again. They had plenty of evidence already that they chose to ignore. And now they chose to further risk the public health.

I looked into the record and saw that Supervisor #7 signed the review at 10:18 am on January 2nd, 2025. It said, "Signing for Christy Leppanen," above her digital signature. My Secondary Reviewer #3 then signed the review at 10:38 am. Findings were then made that same day at 3:35 pm. So, the assessment had been completed and reviewed, to reach findings about the first ever oral pouch products legally sold in the U.S., between 10:38 am and 3:35 pm on January 2nd, 2025.

I then noticed in the publicly posted Decision Summary that I was credited with making findings about those 20 products on January 2nd, 2025. It said, "A finding of no significant impact (FONSI) was signed by Christy Leppanen on January 2, 2025."

What?

I didn't do that.

Those January 2nd, 2025, findings falsely claimed, "The Center for Tobacco Products of the Food and Drug Administration (FDA) has carefully considered the potential environmental effects of these actions and has concluded that these actions will not have significant impacts on the quality of the human environment."

Carefully?

Not even close.

So, on January 23rd, 2025, at 2:24 pm, I alerted who I needed to alert: my Secondary Reviewer #3, and our Supervisor #7, Branch Chief #7, Division Associate Director, Division Director, Office of Science Director #2, and CTP Director #2. I wrote, "I just learned that on page 56 in the publicly posted attachment, I'm listed as the individual who made findings associated with assessments for 20 [product name] oral pouch products: 'A finding of no significant impact (FONSI) was signed by Christy Leppanen on January 2, 2025.'"

I didn't make those findings, and I did not sign the FONSI. So, I wrote, "I did not sign the FONSI." I'd never made findings before and didn't think I even had the authority to make findings. So, I wrote, "I do not have the authority to sign findings and would not have signed the FONSI if asked."

I was on the other side of the Atlantic Ocean, in Germany, from December 31st, 2024, through January 9th, 2025, when those findings were made. While I was outside of the country, where I couldn't take my work computer, the review was finalized, then the assessment for those 20 products was completed and finalized, and then findings were made, all in less than five hours.

On January 24th, 2025, at 3:43 pm, our Division Director replied to all of us and wrote, "Hi Christy, this is confirmation that you are not the signatory of the [oral pouch product name] review and that you did not sign the FONSI. We will correct the public record once public communications are back up and running. Thank you."

Epilogue | Gold Standard Science

JULY 1ST, 2025, CAME AND WENT without those required improvements that the U.S. Office of Special Counsel pointed to as a solution. It's now February 4th, 2026, and public communications have been back up and running for a long time. However, the FDA hasn't corrected those 20 flawed determinations our Division Director said would be corrected. It's been over a year.

And what about all those other flawed records that aren't public?

Further, CTP falsely advertises those 20 oral pouch products in its online public posting, "The Relative Risks of Tobacco Products," as having "undergone rigorous scientific review," without knowing what the pouch was made of.

Rigorous is defined as "scrupulously accurate."

In that online Searchable Tobacco Products Database, CTP maintains at https://www.accessdata.fda.gov/scripts/searchtobacco to inform the public, we still see the same deceptive practices and less transparency, and information we know the agency knows about is still excluded. And CTP has added six more products in "The Relative Risks of Tobacco Products" public posting (https://www.fda.gov/tobacco-products/health-effects-tobacco-use/relative-risks-tobacco-products).

On December 19th, 2025, CTP granted marketing orders for six more oral pouch products. On pages 5, 17, and 27 of the agency's Decision Summary for those six products, CTP admits to considering only 11 of the 93 constituents that the agency is required to publicly list the amounts of, for each brand and subbrand. And CTP again fails to post that information, even for the 11 constituents considered.

Further, the agency again fails to mention consumption during use and focuses on air emissions and littering. Like Secondary Reviewer #3 had said, CTP again decided, "We just have to put some shit down here." Only in its consideration of disposal, is there a hint about what those six pouches are made of. It says on page 10 of the agency's assessment, "The plastic resin comprising the pouch material may potentially persist in the environment."

What kind of "plastic resin"? In what "environment"?

Known evidence is again excluded, for example, from over 85 papers published on January 3rd, 2025, in a review, "Genotoxicity of Microplastics on Living Organisms: Effects on Chromosomes, DNA and Gene Expression." Instead, on page 13 of its Decision Summary, CTP selects only non-clinical data from *five* studies—by tobacco industry scientists—to inform its discussion of genotoxicity.

Then, on page 15 of CTP's Decision Summary, the Technical Project Lead wrote, "I, as TPL, in consultation with CTPs' Office of Science leadership, determined that the new products would not warrant a Toxicology consult (e.g., the added ingredients did not include chemicals with known genotoxicity)."

"Added" ingredients? No Toxicology consult?! What about the toxicological risks and the high scientific bar required by Congress? CTP relies on information from the tobacco industry in its scientific regulatory process that resembles the already questionable EX REQ pathway rather than performing very extensive reviews that the PMTA pathway and *Gold Standard Science* require.

On May 23rd, 2025, President Trump signed Executive Order 14303, "Restoring Gold Standard Science." That Executive Order says, "Over the last 5 years, confidence that scientists act in the best interests of the public has fallen significantly."

It says, "Unfortunately, the Federal Government has contributed to this loss of trust. In several notable cases, executive departments and agencies have used or promoted scientific information in a highly misleading manner." CTP exemplifies an agency that has used or promoted scientific information in a highly misleading manner.

Executive Order 14303 describes policies that must be developed, for instance, for science that informs agency decisions, and for oversight and enforcement to correct scientific information. It repeats the same tenets we've seen repeated but not implemented by Congress and multiple administrations, this time, that agencies must implement Gold Standard Science that is:

(i) reproducible
(ii) transparent
(iii) communicative of error and uncertainty
(iv) collaborative and interdisciplinary
(v) skeptical of its findings and assumptions

(vi) structured for falsifiability of hypotheses

(vii) subject to unbiased peer review

(viii) accepting of negative results as positive outcomes

(ix) without conflicts of interest

It says scientific determinations should use a "weight of scientific evidence" approach and describes that as "an approach to scientific evaluation in which each piece of relevant information is considered based on its quality and relevance, and then transparently integrated with other relevant information to inform the scientific evaluation prior to making a judgment about the scientific evaluation. Quality and relevance determinations, at a minimum, should include consideration of study design, fitness for purpose, replicability, peer review, and transparency and reliability of data."

It says, "Employees' communication of scientific information shall be consistent with the results of the relevant analysis and evaluation and, to the extent that uncertainty is present, the degree of uncertainty should be communicated. Communications involving a scientific model or information derived from a scientific model should include reference to any material assumptions that inform the model's outputs."

It says, "When using scientific information in agency decision making, employees shall transparently acknowledge and document uncertainties, including how uncertainty propagates throughout any models used in the analysis."

It says agencies must make publicly available "the data, analyses, and conclusions associated with scientific and technological information produced or used by the agency" and "models and analyses the agency used to generate such influential scientific information."

It describes things employees should not do, like fabricating data, falsifying data, excluding data limitations, and excluding differences of opinion.

And, it says, "Until the issuance of updated agency scientific integrity policies," that "scientific integrity policies in each agency shall be governed by the scientific integrity policies that existed within the executive branch on January 19, 2021," before improvements required at CTP as an outcome of the wrongdoing disclosure by the U.S. Office of Special Counsel.

Oh no!

Would that mean the 2019 memo directing CTP scientists to "eyeball" data was in play in the interim? Nope, what is outlined in these new policies said all those flawed processes and flawed documents would be corrected. It says, "Within 30 days of the date of this order, the Director of the Office of Science and Technology Policy (OSTP Director) shall, in consultation with the heads of relevant agencies, issue guidance for agencies on implementation of 'Gold Standard Science' in the conduct and management of their respective scientific activities."

It says, "No later than 30 days after the date of this order, agency heads and employees shall adhere to the following rules governing the use, interpretation, and communication of scientific data, unless otherwise provided by law."

It says, "Scientific integrity in the production and use of science by the Federal Government is critical to maintaining the trust of the American people and ensuring confidence in government decisions informed by science."

It says, "My Administration is committed to restoring a gold standard for science to ensure that federally funded research is transparent, rigorous, and impactful, and that Federal decisions are informed by the most credible, reliable, and impartial scientific evidence available."

It says, "Agencies, unless prohibited by law, shall review agency actions taken between January 20, 2021, and January 20, 2025, including regulations, guidance documents, policies, and scientific evaluations and take all appropriate steps, consistent with law, to ensure alignment with the policies and requirements of this order."

That good news was fleshed out further in FDA's August 15th, 2025, implementation of "Gold Standard Science at FDA."

The same tenets of scientific integrity outlined in Executive Order 14303 are described after it says, "No concern is more central to the proper functioning of FDA than the need to ensure that the science underlying agency decisions is sound." And because transparency is part of that policy, we should see things happening!

So, where are we now, in 2026?

Evidence suggests the government isn't interested in Gold Standard Science described in Executive Order 14303 and in FDA's new directives that say, "This policy as a whole ensures the free flow of scientific information and activities, including ensuring that scientists' work and conclusions are accurately represented in agency communications."

Are agency actions consistent with scientific findings?

Are senior leaders with conflicts of interest inappropriately influencing agency decisions?

Questions to Consider

The FDA is required to publish the quantities of 93 harmful and potentially harmful constituents in tobacco products "in each brand and subbrand of tobacco product, in a way that people find understandable and not misleading."

1. How would that information empower citizens to inform their own health decisions?
2. How would the availability of that information to healthcare providers benefit their patients?
3. How would the availability of that information to researchers and regulatory scientists advance science and regulatory science?
4. How would the availability of that information to individuals, healthcare providers, researchers, and regulatory scientists benefit our public health, economy, and wellbeing?
5. Why is that information not made available?
6. Should any additional constituents be considered?

The FDA has failed in its mission to regulate tobacco products.

7. What changes are needed to successfully implement the Family Smoking Prevention and Tobacco Control Act?
8. What other entities might help ensure that success, and how?

Congress has failed in its oversight of the Center for Tobacco Products.

9. What improvements are needed to ensure appropriate oversight?
10. What other entities or mechanisms will ensure appropriate oversight?
11. Outline and describe the effort and resources needed to regulate tobacco products.

Gold Standard Science policies apply across the U.S. Department of Health and Human Services.

12. Describe other examples where Gold Standard Science was not used.
13. In each case, why do you think Gold Standard Science was not used?

References

Blum, Dani (2025) Just How Harmful is Vaping? More Evidence is Emerging. *The New York Times*, July 9th, 2025. https://www.nytimes.com/2025/07/09/well/vaping-health-risks.html

CAKE (1998) "You Turn the Screws." *Prolonging the Magic*, Capricorn Records. https://www.cakemusic.com/music

Carter, Jacob M., Gretchen Goldman, and Charise Johnson (2018) Science under Trump: Voices of Scientists across 16 Federal Agencies (results of 2018 survey). Union of Concerned Scientists, August 7th, 2018. https://www.ucsusa.org/resources/science-under-trump

Centers for Disease Control and Prevention (2020) Tobacco-Related Mortality. U.S. Department of Health and Human Services, April 28th, 2020. https://archive.cdc.gov/www_cdc_gov/tobacco/data_statistics/fact_sheets/health_effects/tobacco_related_mortality/index.htm

Centers for Disease Control and Prevention (2024) Lead Exposure Symptoms and Complications, Childhood Lead Poisoning Prevention. U.S. Department of Health and Human Services, April 10th, 2024. https://www.cdc.gov/lead-prevention/symptoms-complications/index.html

Center for Tobacco Products (2012) Reporting Harmful and Potentially Harmful Constituents in Tobacco Products and Tobacco Smoke under Section 904(a)(3) of the Federal Food, Drug, and Cosmetic Act, Draft Guidance for Industry (Docket Number: FDA-2012-D-0049). Food and Drug Administration, U.S. Department of Health and Human Services. https://www.fda.gov/regulatory-information/search-fda-guidance-documents/reporting-harmful-and-potentially-harmful-constituents-tobacco-products-and-tobacco-smoke-under

Center for Tobacco Products (2015) National Environmental Policy Act; Environmental Assessments for Tobacco Products; Categorical Exclusions, Small Entity Compliance Guide, Guidance for Industry (Docket No. FDA-

2013-N-1282). Food and Drug Administration, U.S. Department of Health and Human Services, 7 pages. https://www.fda.gov/regulatory-information/search-fda-guidance-documents/national-environmental-policy-act-environmental-assessments-tobacco-products-categorical-exclusions

Center for Tobacco Products (2016) Environmental Considerations for Tobacco Product Applications Submitted to CTP, August 2016. Food and Drug Administration, U.S. Department of Health and Human Services, 18 slides. https://fda.report/media/99664/Environmental-Considerations-for-Tobacco-Product-Applications-Submitted-to-CTP-Presentation-Slides.pdf

Center for Tobacco Products (2018) Environmental Tobacco Exposure Workgroup White Paper - January 2018. Food and Drug Administration, U.S. Department of Health and Human Services, 117 pages.

Center for Tobacco Products (2019) Deemed Tobacco Product Applications - A Public Meeting, October 28th-29th, 2019. Food and Drug Administration, U.S. Department of Health and Human Services. https://fda.gov/tobacco-products/ctp-newsroom/deemed-tobacco-product-applications-public-meeting-10282019

Center for Tobacco Products (2023) Premarket Applications: Opportunities for Stakeholder Engagement - A Public Meeting, October 23rd-24th, 2023. Food and Drug Administration, U.S. Department of Health and Human Services. https://fda.gov/tobacco-products/ctp-newsroom/premarket-applications-opportunities-stakeholder-engagement-public-meeting-10232023

Center for Tobacco Products (2024) Programmatic Environmental Assessment for Marketing Orders for Two ENDS Tobacco Products by NJOY, LLC, June 14th, 2024. Food and Drug Administration, U.S. Department of Health and Human Services, 17 pages. https://www.fda.gov/media/179921/download

Center for Tobacco Products (2025) Finding of No Significant Impact, Marketing Orders for ZYN Cool Mint 3 mg, ZYN Cool Mint 6 mg, ZYN Peppermint 3 mg, ZYN Peppermint 6 mg, ZYN Spearmint 3 mg, ZYN Spearmint 6 mg, ZYN Wintergreen 3 mg, ZYN Wintergreen 6 mg, ZYN Citrus 3 mg, ZYN Citrus 6 mg, ZYN Coffee 3 mg, ZYN Coffee 6 mg, ZYN Cinnamon 3 mg, ZYN Cinnamon 6 mg, ZYN Smooth 3mg, ZYN Smooth 6mg, ZYN Chill 3mg, ZYN Chill 6 mg, ZYN Menthol 3 mg, and ZYN Menthol 6 mg, Marketed by Swedish Match USA Inc., January 2, 2025. Food and Drug Administration, U.S. Department of Health and Human Services, 1 page. https://www.accessdata.fda.gov/static/searchtobacco/2-6-25/SMUSA_March_2020_FONSI_PM0000602_JAN_02_2025%20RIF.pdf

Center for Tobacco Products (2025) Programmatic Environmental Assessment for Marketing Orders for New Nicotine Pouch Products Marketed by Helix Innovations LLC, December 16, 2025. Food and Drug Administration, U.S. Department of Health and Human Services, 12 pages. https://www.accessdata.fda.gov/static/searchtobacco/helix/Helix_EA_HelInn_Dec_2024_B1_Redacted.pdf

Center for Tobacco Products (2025) Programmatic Environmental Assessment for Marketing Orders for New Nicotine Pouch Products Marketed by Swedish Match USA Inc., January 2, 2025. Food and Drug Administration, U.S. Department of Health and Human Services, 17 pages. https://www.accessdata.fda.gov/static/searchtobacco/2-6-25/SMUSA_March_2020_ENV_ASSMNT_PM0000602_JAN_02_2025_Redacted.pdf

Center for Tobacco Products (2025) Technical Project Lead (TPL) Review of PMTAs: PM0000593.PD1-PM0000612.PD1, January 16, 2025. Food and Drug Administration, U.S. Department of Health and Human Services, 68 pages. https://www.accessdata.fda.gov/static/searchtobacco/ZYN/PMTA_TPL_PM593-PM612_Zyn_01_13_2025_Redacted.pdf

Center for Tobacco Products (2025) Technical Project Lead (TPL) Review of Nicotine Pouch PMTAs: PM0009121.PD10, PM0009121.PD11, PM0009121.PD13, PM0009121.PD14, PM0009121.PD16, PM0009121.PD17, posted December 19th, 2025. Food and Drug Administration, U.S. Department of Health and Human Services, 42 pages. https://www.fda.gov/media/190306/download?attachment

Center for Tobacco Products (2025) The Relative Risks of Tobacco Products. Food and Drug Administration, U.S. Department of Health and Human Services. https://www.fda.gov/tobacco-products/health-effects-tobacco-use/relative-risks-tobacco-products

Center for Tobacco Products (2025) Tobacco Product Standards. Food and Drug Administration, U.S. Department of Health and Human Services. https://www.fda.gov/tobacco-products/products-guidance-regulations/tobacco-product-standards

Center for Tobacco Products (2026) Searchable Tobacco Products Database. Food and Drug Administration, U.S. Department of Health and Human Services. https://www.accessdata.fda.gov/scripts/searchtobacco

Code of Federal Regulations (2023) 21 C.F.R. § 10.70 https://www.ecfr.gov/current/title-21/chapter-I/subchapter-A/part-10/subpart-B/section-10.70

Collins Dictionary (2025) Definition of *unfair*. https://www.collinsdictionary.com

Combatting the Youth Vaping Epidemic by Enhancing Enforcement against Illegal E-Cigarettes: Hearing before the U.S. Senate Committee on the Judiciary, June 12th, 2024. https://www.judiciary.senate.gov/committee-activity/hearings/combatting-the-youth-vaping-epidemic-by-enhancing-enforcement-against-illegal-e-cigarettes

Desikan, Anita, and Jacob Carter (2023) Getting Science Back on Track: Voices of Scientists across Six Federal Agencies (results of 2022 survey). Union of Concerned Scientists, February 22nd, 2023. https://www.ucsusa.org/resources/scientists-survey-2022

Evaluating FDA Human Foods and Tobacco Programs: Hearing before the U.S. House of Representatives, Committee on Energy and Commerce, Subcommittee on Health, September 10th, 2024. https://energycommerce.house.gov/events/health-subcommittee-hearing-evaluating-fda-human-foods-and-tobacco-programs

Food and Drug Administration (2024) FDA News Release: Youth E-Cigarette Use Drops to Lowest Level in a Decade, Youth Use of Nicotine Pouches Remains Low. U.S. Department of Health and Human Services, September 5th, 2024. https://www.fda.gov/news-events/press-announcements/youth-e-cigarette-use-drops-lowest-level-decade

Food and Drug Administration (2024) Investigation of Elevated Lead & Chromium Levels: Cinnamon Applesauce Pouches (November 2023), FDA Issues Warning Letter and Leverages Import Alerts. U.S. Department of Health and Human Services. https://www.fda.gov/food/outbreaks-foodborne-illness/investigation-elevated-lead-chromium-levels-cinnamon-applesauce-pouches-november-2023

Food and Drug Administration (2024) Microplastics and Nanoplastics in Foods. U.S. Department of Health and Human Services. https://www.fda.gov/food/environmental-contaminants-food/microplastics-and-nanoplastics-foods

Food and Drug Administration (2025) FDA Approves Reformulated Ranitidine Following Comprehensive Safety Review: New Formulation Addresses Previous NDMA Impurity Formation Concerns. U.S. Department of Health and Human Services. https://www.fda.gov/drugs/drug-safety-and-availability/fda-approves-reformulated-ranitidine-following-comprehensive-safety-review

Food and Drug Administration (2025) FDA News Release: FDA Authorizes Marketing of 20 ZYN Nicotine Pouch Products after Extensive Scientific Review. U.S. Department of Health and Human Services, January 16th, 2025. https://www.fda.gov/news-events/press-announcements/fda-authorizes-marketing-20-zyn-nicotine-pouch-products-after-extensive-scientific-review

Food and Drug Administration (2025) Gold Standard Science at FDA. SMG 9001, FDA Staff Manual Guides, Volume IV - Agency Program Directives, General or Multidiscipline. U.S. Department of Health and Human Services, August 15th, 2025. https://www.fda.gov/files/about%20fda/published/9001_internet_08.15.2025.pdf

Global Tobacco and Nicotine Forum (2024) Keynote Address. Global Tobacco and Nicotine Forum 2024, Athens, Greece, September 25th, 2024. https://youtu.be/M8t4wlFXZGk

Goldman, Gretchen, Michael Halpern, Deborah Bailin, Arnold Olali, Charise Johnson, and Tim Donaghy (2015) Progress and Problems: Government Scientists Report on Scientific Integrity at Four Agencies (results of 2015 survey). Union of Concerned Scientists, September 28th, 2015. https://www.ucs.org/resources/progress-and-problems

Jewett, Christina (2024) Teenage E-Cigarette Use Drops to a 10-Year Low. *The New York Times*, September 5th, 2024. https://www.nytimes.com/2024/09/05/health/e-cigarettes-teenagers-vaping.html

Knowles-Carter, Beyoncé Giselle (2024) "Jolene." *Cowboy Carter*, Parkwood Entertainment and Columbia Records. https://music.beyonce.com

Kwong, Emily, Pien Huang, Rachel Carlson, and Rebecca Ramirez (2024) 10 Years after Flint, the Fight to Replace Lead Pipes across the U.S. Continues. *Short Wave*, National Public Radio, April 26th, 2024. https://www.npr.org/2024/04/26/1198909905/flint-water-lead-poisoning-anniversary-chicago

Lu, Wenfeng, Xiaoliang Li, Shuguang Wang, Changli Tu, Lan Qiu, Han Zhang, Chenghui Zhong, Saifeng Li, Yuewei Liu, Jing Liu, and Yun Zhou (2023) New Evidence of Microplastics in the Lower Respiratory Tract: Inhalation through Smoking. *Environmental Science & Technology* 13:8496-8505. https://pubs.acs.org/doi/10.1021/acs.est.3c00716

Merriam-Webster (2025) Definitions of *accurate*, *acknowledge*, *administrative*, *cherry-pick*, *deceptive*, *dirty*, *editorial*, *ethics*, *extensive*, *foul*, *fraud*, *honesty*, *hypothesis*, *illegal*, *integral*, *justice*, *Kafkaesque*, *legal*, *lie*, *nasty*, *rigorous*, *science*, *scientific*, *scientific method*, *state*, *threshold*, *unfair*, *unsportsmanlike*, and *valid*. https://www.merriam-webster.com

NACCHO and Truth Initiative (2024) Zyn and Nicotine Pouches: What Local Tobacco Officials Need to Know. National Association of County and City Health Officials and the Truth Initiative, December 9th, 2024.

Novotny, Thomas E., and Laila Hamzai (2024) Cellulose acetate cigarette filter is hazardous to human health. *Tobacco Control* 33:663-668 https://doi.org/10.1136/tc-2023-057925

Parton, Dolly (1973) "Jolene." *Jolene*, RCA Victor. https://dollyparton.com/album/jolene-album

Reagan-Udall Foundation for the FDA (2022) Operational Evaluation of Certain Components of FDA's Tobacco Program. Reagan-Udall Foundation for the FDA, December 19th, 2022. https://reaganudall.org/projects/operational-evaluation-fdas-tobacco-program

Short, Martin (1985) Interview with Nathan Thurm, a lawyer representing the Tobacco Growers of America, about harms associated with cigarette smoke. Saturday Night News, Season 10, Episode 12, Saturday Night Live, February 2nd, 1985.

Tang, Kuok Ho Daniel (2025) Genotoxicity of Microplastics on Living Organisms: Effects on Chromosomes, DNA and Gene Expression. *Environments* 12(1), 10. https://www.doi.org/10.3390/environments12010010

Trump, Donald J. (2025) Executive Order 14303: Restoring Gold Standard Science. *Federal Register*, May 23rd, 2025. https://www.federalregister.gov/documents/2025/05/29/2025-09802/restoring-gold-standard-science

Truth Initiative (2024) U.S. Retail Sales Data Show 86% of E-cigarette Sales are for Illegal Products. Truth Initiative, November 6th, 2024. https://truthinitiative.org/research-resources/tobacco-industry-marketing/us-retail-sales-data-show-86-e-cigarette-sales-are

Twisted Sister (1984) "We're Not Gonna Take It." *Stay Hungry*, Atlantic Records. https://www.twistedsister.com

Union of Concern Scientists (2008) Survey: FDA Scientists (2006) (results of 2006 survey). Union of Concerned Scientists, July 11th, 2008. https://www.ucs.org/resources/survey-fda-scientists

Union of Concerned Scientists (2010) Voices of Federal Food Safety Scientists and Inspectors (results of 2010 survey). Union of Concerned Scientists, September 13th, 2010. https://www.ucs.org/resources/voices-federal-food-safety-scientists-and-inspectors

Union of Concerned Scientists (2012) Voices of Scientists at the FDA: Measuring Progress on Scientific Integrity (results of 2011 survey). Union of Concerned Scientists, March 8th, 2012. https://www.ucs.org/resources/voices-scientists-fda

U.S. Code (2025) 18 U.S.C. § 1001(a) https://uscode.house.gov/

U.S. Department of Health and Human Services (2025) HHS Organizational Charts. U.S. Department of Health and Human Services. https://www.hhs.gov/about/agencies/orgchart

U.S. Government Accountability Office (2022) Scientific Integrity: HHS Agencies Need to Develop Procedures and Train Staff on Reporting and Addressing Political Interference, Report to Congressional Addressees (GAO-22-104613). U.S. Government Accountability Office, April 20th, 2022. https://www.gao.gov/products/gao-22-104613

U.S. Office of Special Counsel (2022) Unclear Standards for Scientific Review and Stifling of Scientific Dissent Resulted in Flawed FDA Process for Evaluating Potential Harm of Tobacco Products, Disclosure of Wrongdoing. U.S. Office of Special Counsel, December 15th, 2022. https://osc.gov/News/Pages/21-02-FDA-Process-for-Evaluating-Tobacco-Harm.aspx and Whistleblower Response to FDA Report, Whistleblower Proposed Questions for FDA Report, and WB Response to Supplemental Report - OSC File No. DI-20-0372. U.S. Office of Special Counsel, 14 pages. https://osc.gov/Documents/Public%20Files/FY22/DI-20-0372/Whistleblower%20Comments;%20OSC%20File%20No.%20DI-20-0372-No%20Redactions%20Needed.pdf

Wiktionary (2025) Definitions of *florid, flourish, flourishment, lie, lofty, psychosis*, and *rhetorical.* https://www.wiktionary.org

World Health Organization (2025) WHO Study Group on Tobacco Product Regulation. Report on the Scientific Basis of Tobacco Product Regulation: Tenth Report of a WHO Study Group (WHO Technical Report Series, No. 1066). World Health Organization, Geneva. https://www.who.int/publications/b/81861

Xu, Jianchao, Nicholas John Huls, Amber S. Jannasch, and Eric J. Munson (2025) Crystal Defects Cause Nitrosamine Formation in Ranitidine under Accelerated Storage Conditions. *Journal of Pharmaceutical Sciences* 114(12). https://doi.org/10.1016/j.xphs.2025.103960

Referenced information was available at URLs on February 4th, 2026.

About the Author

Christy Leppanen, Ph.D., is an environmental scientist with experience where statutes governing biological, physical, and chemical contaminants are applied. From 2021 to 2025, she was the first microplastic pollution subject matter expert at the FDA Center for Tobacco Products. She was hired to support the agency's compliance with the National Environmental Policy Act, the Family Smoking Prevention and Tobacco Control Act of 2009, and the Federal Food, Drug, and Cosmetic Act.

Dr. Leppanen earned a B.S. in Biology and Chemistry from the University of Memphis, an M.S. in Biology from the University of Memphis, and a Ph.D. in Biological Sciences from the University of Maine. She has worked with NGOs, with state, territorial, and federal governments, with international efforts, and taught and published in environmental toxicology, entomology, invasion biology, and conservation science. Dr. Leppanen was an Associate Editor of the journal *Biological Invasions,* and her work has been supported by the National Science Foundation, the National Fish and Wildlife Foundation, the National Geographic Society, and the University of Washington's Helen R. Whiteley Center.

She also wrote the 2014 Foreword INDIES Bronze Winner for Horror, *Bulletin of ZOMBIE Research: Volume 1,* that was awarded the Kirkus Star.

www.ingramcontent.com/pod-product-compliance
Lightning Source LLC
LaVergne TN
LVHW090313210926
848480LV00001B/228
* 9 7 9 8 9 9 4 3 6 1 6 7 2 *